THE ART OF STATISTICS

Also by David Spiegelhalter

The Norm Chronicles

THE
ART OF
STATISTICS

HOW TO LEARN
FROM DATA

DAVID SPIEGELHALTER

BASIC BOOKS

New York

Basic Books
Hachette Book Group
1290 Avenue of the Americas, New York, NY 10104
www.basicbooks.com

Printed in the United States of America

Originally published in March 2019 by Pelican, an imprint of Penguin Books, in the United Kingdom

First US Edition: September 2019

Published by Basic Books, an imprint of Perseus Books, LLC, a subsidiary of Hachette Book Group, Inc. The Basic Books name and logo is a trademark of the Hachette Book Group.

The Hachette Speakers Bureau provides a wide range of authors for speaking events. To find out more, go to www.hachettespeakersbureau.com or call (866) 376-6591.

The publisher is not responsible for websites (or their content) that are not owned by the publisher.

Print book interior design by Matthew Young.

Library of Congress Cataloging-in-Publication Data has been applied for.

ISBNs: 978-1-5416-1851-0 (hardcover); 978-1-5416-1852-7 (ebook)

LSC-C

10 9 8 7 6 5 4 3 2 1

To statisticians everywhere, with their endearing
traits of pedantry, generosity, integrity,
and desire to use data in the best way possible

Contents

List of Figures

List of Tables

ACKNOWLEDGEMENTS

Any insights gained from a long career in statistics come from listening to inspiring colleagues. These are too numerous even for a statistician to count, but a shortlist of those I have stolen most from might include Nicky Best, Sheila Bird, David Cox, Philip Dawid, Stephen Evans, Andrew Gelman, Tim Harford, Kevin McConway, Wayne Oldford, Sylvia Richardson, Hetan Shah, Adrian Smith and Chris Wild. I am grateful to them and so many others for encouraging me in a challenging subject.

This book has been a long time in development, entirely due to my chronic procrastination. So I would primarily like to thank Laura Stickney of Penguin for not only commissioning the book, but remaining calm as the months, and years, went by, even when the book was finished and we still could not agree on a title. And all credit to Jonathan Pegg for negotiating me a fine deal, Jane Birdsell for showing huge patience when editing, and all the production staff at Penguin for their meticulous work.

I am very grateful for permission to adapt illustrations, specifically Chris Wild (Figure 0.3), James Grime (Figure 2.1), Cath Mercer of Natsal (Figures 2.4 and 2.10), Office for National Statistics (Figures 2.9, 8.5 and 9.4), Public Health England

(Figure 6.7), Paul Barden (Figure 9.2), and BBC (Figure 9.3). UK public sector information is licensed under the Open Government Licence v3.0.

I am not a good R programmer, and Matthew Pearce and Maria Skoularidou helped me enormously in doing the analyses and graphics. I also struggle with writing, and so am indebted to numerous people who read and commented on chapters, including George Farmer, Alex Freeman, Cameron Brick, Michael Posner, Sander van der Linden and Simone Warr: in particular Julian Gilbey had a fine eye for errors and ambiguity.

Above all, I must thank Kate Bull not only for her vital comments on the text, but also for supporting me through times that have been both good (writing in a beach hut in Goa) and not so good (a wet February juggling too many commitments).

I am also profoundly grateful to David and Claudia Harding for both their financial support and their continued encouragement, which has enabled me to do such fun things over the last ten years.

Finally, much as I would like to find someone else to blame, I am afraid I must acknowledge full responsibility for the inevitable remaining inadequacies of this book.

CODE FOR EXAMPLES

R code and data for reproducing most of the analyses and Figures are available from https://github.com/dspiegel29/ArtofStatistics. I am grateful for the assistance received in preparing this material.

Introduction

> The numbers have no way of speaking for themselves. We speak for them. We imbue them with meaning.
>
> — Nate Silver, *The Signal and the Noise*[1]

Why We Need Statistics

Harold Shipman was Britain's most prolific convicted murderer, though he does not fit the archetypal profile of a serial killer. A mild-mannered family doctor working in a suburb of Manchester, between 1975 and 1998 he injected at least 215 of his mostly elderly patients with a massive opiate overdose. He finally made the mistake of forging the will of one of his victims so as to leave him some money: her daughter was a solicitor, suspicions were aroused, and forensic analysis of his computer showed he had been retrospectively changing patient records to make his victims appear sicker than they really were. He was well known as an enthusiastic early adopter of technology, but he was not tech-savvy enough to realize that every change he made was time-stamped (incidentally, a good example of data revealing hidden meaning).

Of his patients who had not been cremated, fifteen were exhumed and lethal levels of diamorphine, the medical form of heroin, were found in their bodies. Shipman was subsequently tried for fifteen murders in 1999, but chose not to offer any defence and never uttered a word at his trial. He was found guilty and jailed for life, and a public inquiry was set up to determine what crimes he might have committed apart from those for which he had been tried, and whether he could have been caught earlier. I was one of a number of statisticians called to give evidence at the public inquiry, which concluded that he had definitely murdered 215 of his patients, and possibly 45 more.[2]

This book will focus on using **statistical science*** to answer the kind of questions that arise when we want to better understand the world – some of these questions will be highlighted in a box. In order to get some insight into Shipman's behaviour, a natural first question is:

> What kind of people did Harold Shipman murder, and when did they die?

The public inquiry provided details of each victim's age, gender and date of death. Figure 0.1 is a fairly sophisticated visualization of this data, showing a scatter-plot of the age of victim against their date of death, with the shading of the points indicating whether the victim was male or female. Bar-charts have been superimposed on the axes showing the pattern of ages (in 5–year bands) and years.

Some conclusions can be drawn by simply taking some

---

* Terms in **bold** appear in the Glossary at the end of the book, which provides both basic and technical definitions.

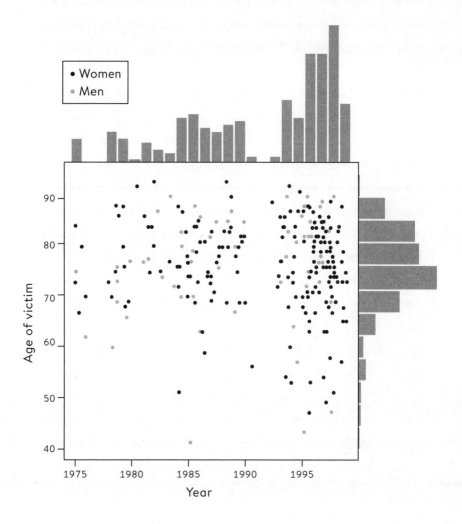

Figure 0.1
A scatter-plot showing the age and the year of death of Harold
Shipman's 215 confirmed victims. Bar-charts have been added on
the axes to reveal the pattern of ages and the pattern of years in
which he committed murders.

time to look at the figure. There are more black than white dots, and so Shipman's victims were mainly women. The bar-chart on the right of the picture shows that most of his victims were in their 70s and 80s, but looking at the scatter of points reveals that although initially they were all elderly, some younger cases crept in as the years went by. The bar-chart at the top clearly shows a gap around 1992 when there were no murders. It turned out that before that time Shipman had been working in a joint practice with other doctors but then, possibly as he felt under suspicion, he left to form a single-handed general practice. After this his activities accelerated, as demonstrated by the top bar-chart.

This analysis of the victims identified by the inquiry raises further questions about the way he committed his murders. Some statistical evidence is provided by data on the time of day of the death of his supposed victims, as recorded on the death certificate. Figure 0.2 is a line graph comparing the times of day that Shipman's patients died to the times that a sample of patients of other local family doctors died. The pattern does not require subtle analysis: the conclusion is sometimes known as 'inter-ocular', since it hits you between the eyes. Shipman's patients tended overwhelmingly to die in the early afternoon.

The data cannot tell us *why* they tended to die at that time, but further investigation revealed that he performed his home visits after lunch, when he was generally alone with his elderly patients. He would offer them an injection that he said was to make them more comfortable, but which was in fact a lethal dose of diamorphine: after a patient had died peacefully in front of him, he would change their medical

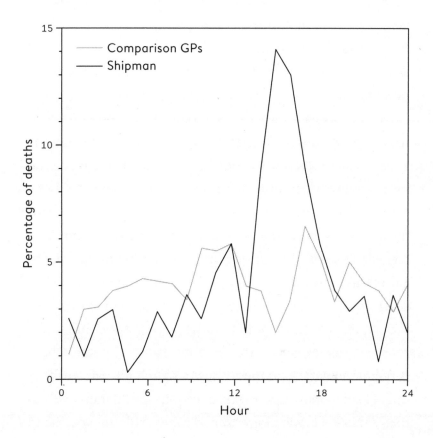

Figure 0.2
The time at which Harold Shipman's patients died, compared to the times at which patients of other local general practitioners died. The pattern does not require sophisticated statistical analysis.

record to make it appear as if this was an expected natural death. Dame Janet Smith, who chaired the public inquiry, later said, 'I still do feel it was unspeakably dreadful, just unspeakable and unthinkable and unimaginable that he should be going about day after day pretending to be this wonderfully caring doctor and having with him in his bag his lethal weapon . . . which he would just take out in the most matter-of-fact way.'

He was taking some risk, since a single post-mortem would have exposed him, but given the age of his patients and the apparent natural causes of death, none were performed. And his reasons for committing these murders have never been explained: he gave no evidence at his trial, never spoke about his misdeeds to anyone, including his family, and committed suicide in prison, conveniently just in time for his wife to collect his pension.

We can think of this type of iterative, exploratory work as 'forensic' statistics, and in this case it was literally true. There is no mathematics, no theory, just a search for patterns that might lead to more interesting questions. The details of Shipman's misdeeds were determined using evidence specific to each individual case, but this data analysis supported a general understanding of how he went about his crimes.

Later in this book, in Chapter 10, we will see whether formal statistical analysis could have helped catch Shipman earlier.* In the meantime, the Shipman story amply demonstrates the great potential of using data to help us understand

---

* Spoiler alert: it almost certainly could have.

the world and make better judgements. This is what statistical science is all about.

Turning the World Into Data

A statistical approach to Harold Shipman's crimes required us to stand back from the long list of individual tragedies for which he was responsible. All those personal, unique details of people's lives, and deaths, had to be reduced to a set of facts and numbers that could be counted and drawn on graphs. This might at first seem cold and dehumanizing, but if we are to use statistical science to illuminate the world, then our daily experiences have to be turned into data, and this means categorizing and labelling events, recording measurements, analysing the results and communicating the conclusions.

Simply categorizing and labelling can, however, present a serious challenge. Take the following basic question, which should be of interest to everyone concerned with our environment:

How many trees are there on the planet?

Before even starting to think about how we might go about answering this question, we first have to settle a rather basic issue. What is a 'tree'? You may feel you know a tree when you see it, but your judgement may differ considerably from others who might consider it a bush or a shrub. So to turn experience into data, we have to start with rigorous definitions.

It turns out that the official definition of a 'tree' is a plant with a woody stem that has a sufficiently large diameter at breast height, known as the DBH. The US Forest Service demands a plant has a DBH of greater than 5 inches (12.7 cm) before officially declaring it a tree, but most authorities use a DBH of 10 cm (4 inches).

But we cannot wander round the entire planet individually measuring each woody-stemmed plant and counting up those that meet this criterion. So the researchers who investigated this question took a more pragmatic approach: they first took a series of areas with a common type of landscape, known as a biome, and counted the average number of trees found per square kilometre. They then used satellite imaging to estimate the total area of the planet covered by each type of biome, carried out some complex statistical modelling, and eventually came up with an estimated total of 3.04 trillion (that is 3,040,000,000,000) trees on the planet. This sounds a lot, except they reckoned there used to be twice this number.[*3]

If authorities differ about what they call a tree, it should be no surprise that more nebulous concepts are even more challenging to pin down. To take an extreme example, the official definition of 'unemployment' in the UK was changed at least thirty-one times between 1979 and 1996.[4] The definition of Gross Domestic Product (GDP) is continually being revised, as when trade in illegal drugs and prostitution was added to the UK GDP in 2014; the estimates used some unusual data

---

* This figure is reported with a margin of error of 0.1 trillion, meaning the researchers were confident the true figure is in a range from 2.94 to 3.14 trillion (I admit feeling this may be rather too accurate, given the many assumptions made in the modelling). They also estimated that 15 billion (15,000,000,000) trees are being cut down each year, and that the planet has lost 46% of its trees since the start of human civilization.

sources – for example Punternet, a review website that rates prostitution services, provided prices for different activities.[5]

Even our most personal feelings can be codified and subjected to statistical analysis. In the year ending September 2017, 150,000 people in the UK were asked as part of a survey: 'Overall, how happy did you feel yesterday?'[6] Their average response, on a scale from zero to ten, was 7.5, an improvement from 2012 when it was 7.3, which might be related to economic recovery since the financial crash of 2008. The lowest scores were reported for those aged between 50 and 54, and the highest between 70 and 74, a typical pattern for the UK.*

Measuring happiness is hard, whereas deciding whether someone is alive or dead should be more straightforward: as the examples in this book will demonstrate, survival and mortality is a common concern of statistical science. But in the US each state can have its own legal definition of death, and although the Uniform Declaration of Death Act was introduced in 1981 to try to establish a common model, some small differences remain. Someone who had been declared dead in Alabama could, at least in principle, cease to be legally dead were they across the state border in Florida, where the registration must be made by two qualified doctors.[7]

These examples show that statistics are always to some extent constructed on the basis of judgements, and it would be an obvious delusion to think the full complexity of personal experience can be unambiguously coded and put into a spreadsheet or other software. Challenging though it is to

---

* Which, if I were average, would give me something to look forward to.

define, count and measure characteristics of ourselves and the world around us, it is still just information, and only the starting point to real understanding of the world.

Data has two main limitations as a source of such knowledge. First, it is almost always an imperfect measure of what we are really interested in: asking how happy people were last week on a scale from zero to ten hardly encapsulates the emotional wellbeing of the nation. Second, anything we choose to measure will differ from place to place, from person to person, from time to time, and the problem is to extract meaningful insights from all this apparently random **variability**.

For centuries, statistical science has faced up to these twin challenges, and played a leading role in scientific attempts to understand the world. It has provided the basis for interpreting data, which is always imperfect, in order to distinguish important relationships from the background variability that makes us all unique. But the world is always changing, as new questions are asked and new sources of data become available, and statistical science has had to change too.

People have always counted and measured, but modern statistics as a discipline really began in the 1650s when, as we shall see in Chapter 8, probability was properly understood for the first time by Blaise Pascal and Pierre de Fermat. Given this solid mathematical basis for dealing with variability, progress was then remarkably rapid. When combined with data on the ages at which people die, the theory of probability provided a firm basis for calculating pensions and annuities. Astronomy was revolutionized when scientists grasped how probability theory could handle variability in measurements.

Victorian enthusiasts became obsessed with collecting data about the human body (and everything else), and established a strong connection between statistical analysis and genetics, biology and medicine. Then in the twentieth century statistics became more mathematical and, unfortunately for many students and practitioners, the topic became synonymous with the mechanical application of a bag of statistical tools, many named after eccentric and argumentative statisticians that we shall meet later in this book.

This common view of statistics as a basic 'bag of tools' is now facing major challenges. First, we are in an age of **data science**, in which large and complex data sets are collected from routine sources such as traffic monitors, social media posts and internet purchases, and used as a basis for technological innovations such as optimizing travel routes, targeted advertising or purchase recommendation systems – we shall look at **algorithms** based on **'big data'** in Chapter 6. Statistical training is increasingly seen as just one necessary component of being a data scientist, together with skills in data management, programming and algorithm development, as well as proper knowledge of the subject matter.

Another challenge to the traditional view of statistics comes from the huge rise in the amount of scientific research being carried out, particularly in the biomedical and social sciences, combined with pressure to publish in high-ranking journals. This has led to doubts about the reliability of parts of the scientific literature, with claims that many 'discoveries' cannot be reproduced by other researchers – such as the continuing dispute over whether adopting an assertive posture popularly known as a 'power pose' can induce hormonal and other changes.[8] The inappropriate use of standard statistical methods has received a fair share of the blame for what has become known as the reproducibility or replication crisis in science.

With the growing availability of massive data sets and user-friendly analysis software, it might be thought that there is less need for training in statistical methods. This would be naïve in the extreme. Far from freeing us from the need for statistical skills, bigger data and the rise in the number and complexity of scientific studies makes it even more difficult to draw appropriate conclusions. More data means that we need to be even more aware of what the evidence is actually worth.

For example, intensive analysis of data sets derived from routine data can increase the possibility of false discoveries, both due to systematic bias inherent in the data sources and from carrying out many analyses and only reporting whatever looks most interesting, a practice sometimes known as 'data-dredging'. In order to be able to critique published scientific work, and even more the media reports which we all encounter on a daily basis, we should have an acute awareness of the dangers of selective reporting, the need for scientific claims to be replicated by independent researchers, and the danger of over-interpreting a single study out of context.

All these insights can be brought together under the term **data literacy**, which describes the ability to not only carry out statistical analysis on real-world problems, but also to understand and critique any conclusions drawn by others on the basis of statistics. But improving data literacy means changing the way statistics is taught.

Teaching Statistics

Generations of students have suffered through dry statistics courses based on learning a set of techniques to be applied in different situations, with more regard to mathematical theory than understanding both why the formulae are being used, and the challenges that arise when trying to use data to answer questions.

Fortunately this is changing. The needs of data science and data literacy demand a more problem-driven approach, in which the application of specific statistical tools is seen as just one component of a complete cycle of investigation. The **PPDAC** structure has been suggested as a way of representing a problem-solving cycle, which we shall adopt throughout this book.[9] Figure 0.3 is based on an example from New Zealand, which has been a world-leader in statistics education in schools.

The first stage of the cycle is specifying a Problem; statistical inquiry always starts with a question, such as our asking about the pattern of Harold Shipman's murders or the number of trees in the world. Later in this book we shall focus on problems ranging from the expected benefit of different therapies immediately following breast cancer surgery, to why old men have big ears.

It is tempting to skip over the need for a careful Plan. The Shipman question simply required the collection of as much data as possible on his victims. But the people counting trees paid meticulous attention to precise definitions and how to carry out the measurements, since confident conclusions can only be drawn from a study which has been appropriately

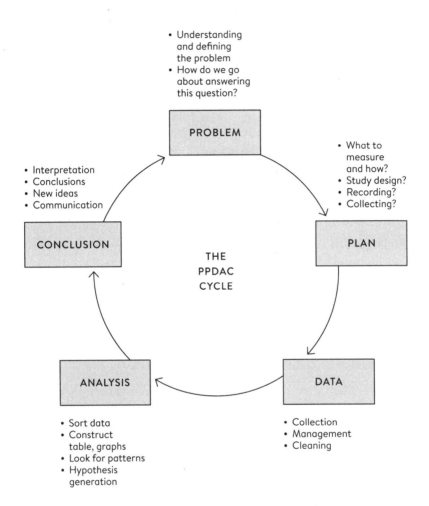

- Understanding and defining the problem
- How do we go about answering this question?

PROBLEM

- What to measure and how?
- Study design?
- Recording?
- Collecting?

- Interpretation
- Conclusions
- New ideas
- Communication

CONCLUSION

THE PPDAC CYCLE

PLAN

ANALYSIS

DATA

- Sort data
- Construct table, graphs
- Look for patterns
- Hypothesis generation

- Collection
- Management
- Cleaning

Figure 0.3

The PPDAC problem-solving cycle, going from Problem, Plan, Data, Analysis to Conclusion and communication, and starting again on another cycle.

designed. Unfortunately, in the rush to get data and start analysis, attention to design is often glossed over.

Collecting good Data requires the kind of organizational and coding skills that are being seen as increasingly important in data science, particularly as data from routine sources may need a lot of cleaning in order to get it ready to be analysed. Data collection systems may have changed over time, there may be obvious errors, and so on – the phrase 'found data' neatly communicates that it may be rather messy, like something picked up in the street.

The Analysis stage has traditionally been the main emphasis of statistics courses, and we shall cover a range of analytic techniques in this book; but sometimes all that is required is a useful visualization, as in Figure 0.1. Finally, the key to good statistical science is drawing appropriate Conclusions that fully acknowledge the limitations in the evidence, and communicating them clearly, as in the graphical illustrations of the Shipman data. Any conclusions generally raise more questions, and so the cycle starts over again, as when we started looking at the time of day when Shipman's patients died.

Although in practice the PPDAC cycle laid out in Figure 0.3 may not be followed precisely, it underscores that formal techniques for statistical analysis play only one part in the work of a statistician or data scientist. Statistical science is a lot more than a branch of mathematics involving esoteric formulae with which generations of students have (often reluctantly) struggled.

This Book

When I was a student in Britain in the 1970s, there were just three TV channels, computers were the size of a double wardrobe, and the closest thing we had to Wikipedia was on the imaginary handheld device in Douglas Adams' (remarkably prescient) *Hitchhiker's Guide to the Galaxy*. For self-improvement we therefore turned to Pelican books, and their iconic blue spines were a standard feature of every student bookshelf.

Because I was studying statistics, my Pelican collection featured *Facts from Figures* by M. J. Moroney (1951) and *How to Lie with Statistics* by Darrell Huff (1954). These venerable publications sold in the hundreds of thousands, reflecting both the level of interest in statistics and the dismal lack of choice at that time. These classics have stood up remarkably well to the intervening sixty-five years, but the current era demands a different approach to teaching statistics based on the principles laid out above.

This book therefore uses real-world problem-solving as a starting point for introducing statistical ideas. Some of these ideas may seem obvious, but some are more subtle and may require some mental effort, although mathematical skills will not be needed. Compared to traditional texts, this book focuses on conceptual issues rather than technicalities, and features only a few, fairly innocuous equations supported by a Glossary. Software is a vital part of any work in data science and statistics but it is not a focus of this book – tutorials are readily available for freely available environments such as R and Python.

The questions featured in the boxes can all, to a certain

extent, be answered through statistical analysis, although they differ widely in their scope. Some are important scientific hypotheses, such as whether the Higgs boson exists, or if there really is convincing evidence for extra-sensory perception (ESP). Others are questions about health care, such as whether busier hospitals have higher survival rates, and if screening for ovarian cancer is beneficial. Sometimes we just want to estimate quantities, such as the cancer risk from bacon sandwiches, the number of sexual partners people in Britain have in their lifetime, and the benefit of taking a daily statin.

And some questions are just interesting, such as identifying the luckiest survivor from the *Titanic*; whether Harold Shipman could have been caught earlier; and assessing the probability that a skeleton found in a Leicester car park really was that of Richard III.

This book is intended for both students of statistics who are seeking a non-technical introduction to the basic issues, and general readers who want to be more informed about the statistics they encounter both in their work and in everyday life. My emphasis is on handling statistics skilfully and with care: numbers may appear to be cold, hard facts, but the attempts to measure trees, happiness and death have already shown that they need to be treated with delicacy.

Statistics can bring clarity and insight into the problems we face, but we are all familiar with the way they can be abused, often to promote an opinion or simply to attract attention. The ability to assess the trustworthiness of statistical claims seems a key skill in the modern world, and I hope that this book may help to empower people to question the numbers that they encounter in their daily life.

Summary

- Turning experiences into data is not straightforward, and data is inevitably limited in its capacity to describe the world.
- Statistical science has a long and successful history, but is now changing in the light of increased availability of data.
- Skill in statistical methods plays an important part of being a data scientist.
- Teaching statistics is changing from a focus on mathematical methods to one based on an entire problem-solving cycle.
- The PPDAC cycle provides a convenient framework: Problem – Plan – Data – Analysis – Conclusion and communication.
- Data literacy is a key skill for the modern world.

CHAPTER 1

Getting Things in Proportion: Categorical Data and Percentages

What happened to children having heart surgery in Bristol between 1984 and 1995?

Joshua L was 16 months old and had transposition of the great arteries, a severe form of congenital heart disease in which the main vessels coming from the heart are attached to the wrong ventricle. He needed an operation to 'switch' the arteries, and just after 7 a.m. on 12 January 1995 his parents said goodbye to him and watched as he was taken for his surgery in Bristol Royal Infirmary. But Joshua's parents were unaware that stories about the poor surgical survival rates at Bristol had been circulating since the early 1990s. Nobody told them that nurses had left the unit rather than continue telling parents that their child had died, or that the previous evening there had been a late-night meeting at which it had been debated whether to cancel Joshua's operation.[1]

Joshua died on the operating table. The following year the General Medical Council (the medical regulator) launched an investigation after complaints from Joshua's and other bereaved parents, and in 1998 two surgeons and the ex-chief executive were found guilty of serious medical misconduct.

Public concern did not die down, and an official inquiry was ordered: this brought in a team of statisticians who were given the grim task of comparing the survival rates in Bristol with elsewhere in the UK between 1984 and 1995. I led this team.

We first had to determine how many children had had heart surgery, and how many had died. This sounds like it should be straightforward but, as shown in the previous chapter, simply counting events can be challenging. What is a 'child'? What counts as 'heart surgery'? When can death be attributed to surgery? And even when these definitions have been decided, could we determine how many of each there had been?

We took a 'child' as anyone under 16, and focused on 'open' surgery in which the heart had been stopped and its function replaced by cardio-pulmonary bypass. There can be multiple operations per admission, but these were considered as one event. Deaths were counted if they occurred within 30 days of the operation, whether or not in hospital or due to the surgery. We knew that death was an imperfect measure of the quality of the outcome, as it ignored children who were brain-damaged or otherwise disabled as a result of the surgery, but we did not have the data on longer-term outcomes.

The main source of data was national Hospital Episode Statistics (HES), which were derived from administrative data entered by low-paid coders. HES had a poor reputation among doctors, but this source had the great advantage that it could be linked to national death records. There was also a parallel system of data submitted directly to a Cardiac Surgical Registry (CSR) established by the surgeons' professional society.

These two sources of data, though they were supposed to be about exactly the same practice, showed considerable disagreement: for 1991–1995, HES said there had been 62 deaths out of 505 open operations (14%), whereas CSR said there had been 71 deaths out of 563 operations (13%). No less than five additional local sources of data were available, from anaesthetic records to the surgeons' own personal logs. Bristol was awash with data, but none of the data sources could be considered the 'truth', and nobody had taken responsibility for analysing and acting on the surgical outcomes.

We calculated that if patients at Bristol had the average risk prevailing elsewhere in the UK, Bristol would have expected to have had 32 deaths over this period, instead of the 62 recorded in HES, which we reported as '30 excess deaths' between 1991 and 1995.* The exact numbers varied according to the data sources, and it may seem extraordinary that we could not even establish the basic facts about the number of operations and their outcome, although current record systems should be better.

These findings had wide press coverage, and the Bristol inquiry led to a major change in attitudes to monitoring clinical performance: no longer was the medical profession trusted to police itself. Mechanisms to publicly report hospital survival data were established, although, as we shall now see, the way in which that data is displayed can itself influence the perception of audiences.

---

* I now regret using the term 'excess deaths', since newspapers later interpreted this as meaning 'avoidable deaths'. But around half of hospitals will have more deaths than expected just by chance alone, and only a few of these deaths might be judged avoidable.

Communicating Counts and Proportions

Data that records whether individual events have happened or not is known as **binary data,** as it can only take on two values, generally labelled as yes and no. Sets of binary data can be summarized by the number of times and the percentage of cases in which an event occurred.

The theme of this chapter is that the basic presentation of statistics is important. In a sense we are jumping to the last step of the PPDAC cycle in which conclusions are communicated, and while the form of this communication has not traditionally been considered an important topic in statistics, rising interest in data visualization reflects a change in this attitude. So both in this chapter and the next we shall concentrate on ways of displaying data so that we can quickly get the gist of what is going on without detailed analysis, starting with a look at alternative ways of displaying data that, largely because of the Bristol inquiry, are now publicly available.

Table 1.1 shows the outcomes of nearly 13,000 children who had heart surgery in the UK and Ireland between 2012 and 2015.[2] Two hundred and sixty-three babies died within 30 days of their operation, and every one of these deaths is a tragedy to the family involved. It will be little consolation to them that survival rates have improved hugely from the time of the Bristol inquiry, and now average 98%, and so there is a more hopeful prospect for families of children facing heart surgery.

A table can be considered as a type of graphic, and requires careful design choices of colour, font and language to ensure engagement and readability. The audience's emotional response to the table may also be influenced by the choice

Hospital	Number of babies having surgery	Number surviving for at least 30 days after surgery	Number dying within 30 days of surgery	Percentage surviving	Percentage dying
London, Harley Street	418	413	5	98.8	1.2
Leicester	607	593	14	97.7	2.3
Newcastle	668	653	15	97.8	2.2
Glasgow	760	733	27	96.3	3.7
Southampton	829	815	14	98.3	1.7
Bristol	835	821	14	98.3	1.7
Dublin	983	960	23	97.7	2.3
Leeds	1,038	1,016	22	97.9	2.1
London, Brompton	1,094	1,075	19	98.3	1.7
Liverpool	1,132	1,112	20	98.2	1.8
London, Evelina	1,220	1,185	35	97.1	2.9
Birmingham	1,457	1,421	36	97.5	2.5
London, Great Ormond Street	1,892	1,873	19	99.0	1.0
Total	12,933	12,670	263	98.0	2.0

Table 1.1
Outcomes of children's heart surgery in UK and Irish hospitals between 2012 and 2015, in terms of survival or not, 30 days after surgery.

of which columns to display. Table 1.1 shows the results in terms of both survivors and deaths, but in the US *mortality* rates from child heart surgery are reported, while the UK provides *survival* rates. This is known as negative or positive **framing**, and its overall effect on how we feel is intuitive and well-documented: '5% mortality' sounds worse than '95% survival'. Reporting the actual number of deaths as well as the percentage can also increase the impression of risk, as this total might then be imagined as a crowd of real people.

A classic example of how alternative framing can change the emotional impact of a number is an advertisement that appeared on the London Underground in 2011, proclaiming that '99% of young Londoners do not commit serious youth violence'. These ads were presumably intended to reassure passengers about their city, but we could reverse its emotional impact with two simple changes. First, the statement means that 1% of young Londoners *do* commit serious violence. Second, since the population of London is around 9 million, there are around 1 million people aged between 15 and 25, and if we consider these as 'young', this means there are 1% of 1 million or a total of 10,000 seriously violent young people in the city. This does not sound at all reassuring. Note the two tricks used to manipulate the impact of this statistic: convert from a positive to a negative frame, and then turn a percentage into actual numbers of people.

Ideally both positive and negative frames should be presented if we want to provide impartial information, although the order of columns might still influence how the table is interpreted. The order of the rows of a table also needs to be considered carefully. Table 1.1 shows the hospitals in order of the number of operations in each, but if they had been presented, say, in order of mortality rates with the highest at the top of the table, this might give the impression that this was a valid and important way of comparing hospitals. Such league tables are favoured by the media and even some politicians, but can be grossly misleading: not only because the differences could be due to chance variation, but because the hospitals may be taking in very different types of cases. In Table 1.1, for example, we might suspect that Birmingham, one of the biggest and most well-known children's hospitals, takes on the most severe cases, and so it would be unfair, to put it mildly, to highlight their apparently unimpressive overall survival rates.*

The survival rates can be presented in a horizontal bar-chart such as the one shown in Figure 1.1. A crucial choice is where to start the horizontal axis: if the values start from 0%, all the bars will be almost the full length of the graphic, which will clearly show the extraordinarily high survival rates, but the lines will be indistinguishable. But the oldest trick of misleading graphics is to start the axis at say 95%, which will make the hospitals look extremely different, even if the variation is in fact only what is attributable to chance alone.

Choosing the start of the axis therefore presents a dilemma. Alberto Cairo, author of influential books on data visualization,[3] suggests you should always begin with a 'logical and meaningful baseline', which in this situation appears difficult to identify – my rather arbitrary choice of 86% roughly

---

* It turns out there is no good evidence for any systematic differences between these hospitals, once the severity of their cases is taken into account[2].

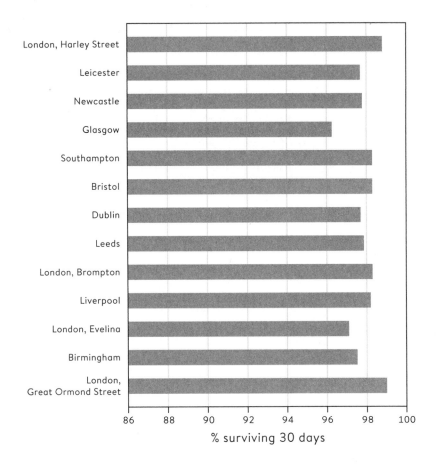

Figure 1.1
Horizontal bar-chart of 30-day survival rates for thirteen
hospitals. The choice of the start of the horizontal axis, here 86%,
can have a crucial effect on the impression given by the graphic.
If the axis starts at 0%, all the hospitals will look indistinguishable,
whereas if we started at 95% the differences would look
misleadingly dramatic.

represents the unacceptably low survival in Bristol twenty years previously.

I began this book with a quotation from Nate Silver, the founder of data-based platform *FiveThirtyEight* and first famous for accurately predicting the 2008 US presidential election, who eloquently expressed the idea that numbers do not speak for themselves – we are responsible for giving them meaning. This implies that communication is a key part of the problem-solving cycle, and I have shown in this section how the message from a set of simple proportions can be influenced by our choices of presentation.

We now need to introduce an important and convenient concept that will help us get beyond simple yes/no questions.

Categorical Variables

A variable is defined as any measurement that can take on different values in different circumstances; it's a very useful shorthand term for all the types of observations that comprise data. Binary variables are yes/no questions such as whether someone is alive or dead and whether they are female or not: both of these vary between people, and can, even for gender, vary within people at different times. **Categorical variables** are measures that can take on two or more categories, which may be

- Unordered categories: such as a person's country of origin, the colour of a car, or the hospital in which an operation takes place.
- Ordered categories: such as the rank of military personnel.

- Numbers that have been grouped: such as levels of obesity, which is often defined in terms of thresholds for the body mass index (BMI).*

When it comes to presenting categorical data, pie charts allow an impression of the size of each category relative to the whole pie, but are often visually confusing, especially if they attempt to show too many categories in the same chart, or use a three-dimensional representation that distorts areas. Figure 1.2 shows a fairly hideous example modelled on the kind offered by Microsoft Excel, showing the proportions of the 12,933 child heart patients from Table 1.1 that are treated in each hospital.

Multiple pie charts are generally not a good idea, as comparisons are hampered by the difficulty in assessing the relative sizes of areas of different shapes. Comparisons are better based on height or length alone in a bar chart. Figure 1.3 shows a simpler, clearer example of a horizontal bar chart of the proportions being treated in each hospital.

Comparing a Pair of Proportions

We have seen how a set of proportions can be elegantly compared using a bar chart, and so it would be reasonable to think that comparing two proportions would be a trivial matter. But when these proportions represent estimates of

---

* The body mass index was developed by Belgian statistician Adolphe Quetelet before 1850, and is defined as BMI = weight (kg)/ height (m)^2. Many different groupings of his index are in use, and current UK definitions for obesity are: Underweight (BMI < 18.5 kg/m^2); Normal (BMI between 18.5 and 25); Overweight (between 25 and 30), Obese (30 to 35) and Morbidly Obese (above 35).

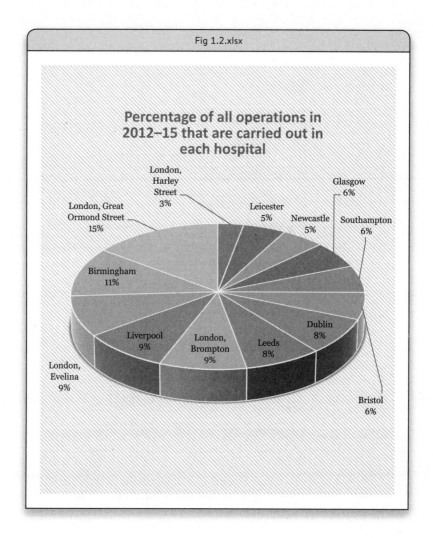

Figure 1.2

The proportion of all child heart operations being carried out in each hospital, displayed in a 3D pie chart from Excel. This deeply unpleasant chart makes categories near the front look bigger, and so makes it impossible to make visual comparisons between hospitals.

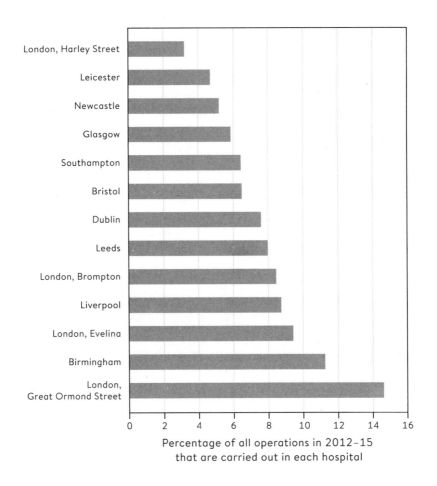

Figure 1.3
Percentage of all child heart operations being carried out in each hospital: a clearer representation using a horizontal bar chart.

the risks of experiencing some harm, then the way in which those risks are compared becomes a serious and contested issue. Here is a typical question:

> What's the cancer risk from bacon sandwiches?

We're all familiar with hyperbolic media headlines that warn us that something mundane increases the risk of some dread occurrence: I like to call these 'cats cause cancer' stories. For example, in November 2015 the World Health Organization's International Agency for Research in Cancer (IARC) announced that processed meat was a 'Group I carcinogen', putting it in the same category as cigarettes and asbestos. This inevitably led to panicky headlines such as the *Daily Record*'s claim that 'Bacon, Ham and Sausages Have the Same Cancer Risk as Cigarettes Warn Experts'.[4]

The IARC tried to quell the fuss by emphasizing that the Group 1 classification was about being confident that an increased risk of cancer existed at all, and said nothing about the actual magnitude of the risk. Lower down in the press release, the IARC reported that 50g of processed meat a day was associated with an increased risk of bowel cancer of 18%. This sounds worrying, but should it be?

The figure of 18% is known as a **relative risk** since it represents the increase in risk of getting bowel cancer between a group of people who eat 50g of processed meat a day, which could, for example, represent a daily two-rasher bacon sandwich, and a group who don't. Statistical commentators took this relative risk and reframed it into a change in **absolute risk**, which means the change in the actual proportion in each group who would be expected to suffer the adverse event.

They concluded that, in the normal run of things, around 6 in every 100 people who do not eat bacon daily would be expected to get bowel cancer in their lifetime. If 100 similar people ate a bacon sandwich every single day of their lives, then according to the IARC report we would expect that 18% more would get bowel cancer, which means a rise from 6 to 7 cases out of 100.* That is one extra case of bowel cancer in all those 100 lifetime bacon-eaters, which does not sound so impressive as the relative risk (an 18% increase), and might serve to put this hazard into perspective. We need to distinguish what is actually dangerous from what sounds frightening.[5]

This bacon sandwich example illustrates the advantage of communicating risks using **expected frequencies**: instead of discussing percentages or probabilities, we just ask, 'What does this mean for 100 (or 1,000) people?' Psychological studies have shown that this technique improves understanding: in fact communicating only that this additional meat-eating led to an '18% increased risk' could be considered manipulative, since we know this phrasing gives an exaggerated impression of the importance of the hazard.[6] Figure 1.4 uses **icon arrays** to directly represent the expected frequencies of bowel cancer in 100 people.

In Figure 1.4 the 'cancer' icons are randomly scattered among the 100. While such scatter has been shown to

---

* Strictly speaking, an 18% relative increase over 6% is 6% × 1.18 = 7.08%, but rounding to 7% is good enough for this level of communication.

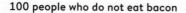

100 people who do not eat bacon

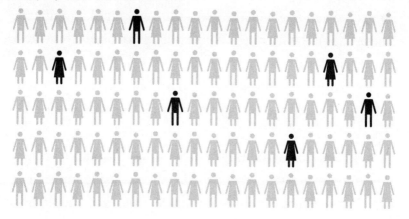

100 people who eat bacon every day

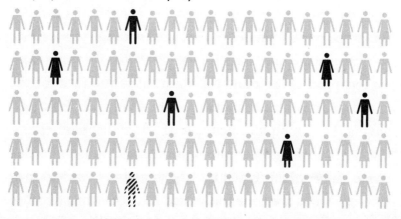

Figure 1.4

Bacon sandwich example using a pair of icon arrays, with randomly scattered icons showing the incremental risk of eating bacon every day. Of 100 people who do not eat bacon, 6 (solid icons) develop bowel cancer in the normal run of events. Of 100 people who eat bacon every day of their lives, there is 1 additional (striped) case.*

---

* Strictly speaking the six solid icons should be positioned differently in the two graphics, as they represent two different groups of 100 people. But this would make the two groups more difficult to compare.

increase the impression of unpredictability, it should only be used when there is a single additional highlighted icon. There should be no need to count icons in order to make a quick visual comparison.

Yet more ways to compare two proportions are shown in Table 1.2, illustrated by the risks for people who do and do not eat bacon.

'1 in X' is a common way of expressing risk, such as saying '1 in 16 people' to represent a 6% risk. But using multiple '1 in . . .' statements is not recommended, as many people find them difficult to compare. For example, when asked the question, 'Which is the bigger risk, 1 in 100, 1 in 10 or 1 in 1,000?', around a quarter of people answered incorrectly: the problem is that the bigger number is associated with the smaller risk, and so some mental dexterity is required to keep things clear.

Technically, the **odds** for an event is the ratio of the chance of the event happening to the chance of it not happening. For example, since, out of 100 non-bacon eaters, 6 will get bowel cancer and 94 won't, the odds of getting bowel cancer in this group is 6/94, sometimes referred to as '6 to 94'. Odds are commonly used in UK betting, but they are also used extensively in statistical modelling of proportions, and this means that medical research commonly expresses the effects associated with treatments or behaviour in terms of **odds ratios**.

Although extremely common in the research literature, odds ratios are a rather unintuitive way to summarize differences in risk. If the events are fairly rare then the odds ratios will be numerically close to the relative risks, as in the case of bacon sandwiches, but for common events the odds ratio

GETTING THINGS IN PROPORTION

Method	Non-bacon eaters	Daily bacon eaters
Event rate	6%	7%
Expected frequency	6 out of 100	7 out of 100
	1 in 16	1 in 14
Odds	6/94	7/93

Comparative measures

Absolute risk difference	1%, or 1 out of 100
Relative risk	1.18, or an 18% increase
'Number Needed to Treat'	100
Odds ratio	(7/93) / (6/94) = 1.18

Table 1.2
Examples of methods for communicating the lifetime risk of bowel cancer with and without a daily bacon sandwich. The 'Number Needed to Treat' is the number of people needing to eat a bacon sandwich every day of their lives, in order to expect one extra case of bowel cancer (and so would perhaps better be defined as the Number Needed to Eat).

can be very different from the relative risk, and the following example shows this can be very confusing for journalists (and others).

> How can a rise from 85% to 87% be called a 20% increase?

Statins are widely taken to reduce cholesterol and the risk of heart attacks and strokes, but some doctors have expressed concern about side effects. A study published in 2013 found that 87% of people taking statins reported muscle pains, compared to 85% in those who did not take statins. Looking at the options for comparing risks shown in Table 1.2, we might report either a 2% increase in absolute risk, or a relative risk of 0.87/0.85 = 1.02, that is a 2% relative increase in risk. The odds in the two groups are given by 0.87/0.13 = 6.7 and 0.85/0.15 = 5.7, and so the odds ratio is therefore 6.7/5.7 = 1.18: exactly the same as for bacon sandwiches, but based on very different absolute risks.

The *Daily Mail* misinterpreted this odds ratio of 1.18 as a relative risk, and produced a headline claiming statins 'raises risk by up to 20 per cent', which is a serious misrepresentation of what the study actually found. But not all the blame can be placed on the journalists: the abstract of the paper mentioned only the odds ratio without mentioning that this corresponded to a difference between absolute risks of 85% vs 87%.[7]

This highlights the danger of using odds ratios in anything but a scientific context, and the advantage of always reporting absolute risks as the quantity that is relevant for an audience, whether they are concerned with bacon, statins or anything else.

The examples in this chapter have demonstrated how the apparently simple task of calculating and communicating proportions can become a complex matter. It needs to be carried out with care and awareness, and the impact of numerical or graphical data summaries can be explored by working with psychologists who are skilled in evaluating the perception of alternative formats. Communication is an important part of the problem-solving cycle, and should not be just a matter of personal preference.

Summary

- Binary variables are yes/no questions, sets of which can be summarized as proportions.
- Positive or negative framing of proportions can change their emotional impact.
- Relative risks tend to convey an exaggerated importance, and absolute risks should be provided for clarity.
- Expected frequencies promote understanding and an appropriate sense of importance.
- Odds ratios arise from scientific studies but should not be used for general communication.
- Graphics need to be chosen with care and awareness of their impact.

CHAPTER 2

Summarizing and Communicating Numbers. Lots of Numbers

Can we trust the wisdom of crowds?

In 1907 Francis Galton, cousin of Charles Darwin and poly-math originator of identification using fingerprints, weather forecasts and eugenics,* wrote a letter to the prestigious science journal *Nature* about his visit to the Fat Stock and Poultry Exhibition in the port city of Plymouth. There he saw a large ox displayed and contestants paying sixpence to guess the 'dressed' weight of the resulting meat after the poor beast had been slaughtered. He got hold of 787 of the tickets that had been filled out and chose the middle value of 1,207 lb (547 kg) as the democratic choice, 'every other estimate being condemned as too high or too low by the majority of voters'. The dressed weight turned out to be 1,198 lb (543 kg), which was remarkably close to his choice based on the 787

---

* Eugenics is the idea that the human race can be improved by selective breeding, either by encouraging the 'fit' to produce more children by, for example, financial incentives, or by preventing the 'unfit' from reproducing, say by encouraging sterilization. Many of the early developers of statistical techniques were enthusiastic eugenicists. The experiences in Nazi Germany put an end to the movement, although the academic journal *Annals of Eugenics* only changed its name to the current *Annals of Genetics* in 1955.

votes.[1] Galton titled his letter 'Vox Populi' (voice of the people), but this process of decision-making is now better known as the **wisdom of crowds.**

Galton carried out what we might now call a data summary: he took a mass of numbers written on tickets and reduced them to a single estimated weight of 1,207 lb. In this chapter we look at the techniques that have been developed in the subsequent century for summarizing and communicating the piles of data that have become available. We will see that numerical summaries of location, spread, trend and correlation are intimately related to how the data can be plotted on paper or a screen. And we shall look at the gentle transition between simply describing the data, and seeking to tell a story through an infographic.

First we will begin with my own attempt at a wisdom-of-crowds experiment, which demonstrates many of the problems that crop up when the real, undisciplined world, with all its capacity for oddity and error, is used as a source of data.

Statistics is not only concerned with serious events such as cancer and surgery. In a rather trivial experiment, mathematics communicator James Grime and I posted a video on YouTube and asked anyone watching to guess the number of jelly beans in a jar. You might want to try this exercise yourself when you see the photo in Figure 2.1 (the true number will be revealed later). Nine hundred and fifteen people provided their guesses, which ranged from 219 to 31,337, and in this chapter we shall look at how such variables can be depicted graphically and summarized numerically.

To start, Figure 2.2 shows three ways of presenting the

Figure 2.1

How many jelly beans are in this jar? We asked this on a YouTube video and got 915 responses. The answer will be given later.

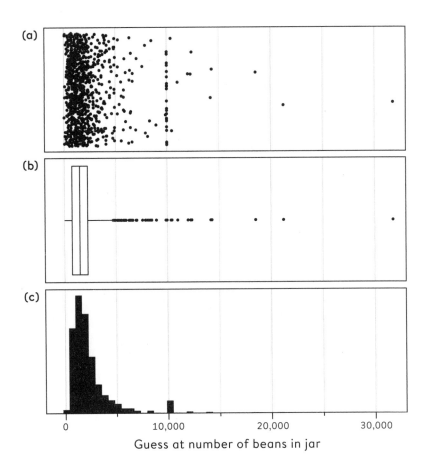

Figure 2.2
Different ways of showing the pattern of 915 guesses of the
number of jelly beans in the jar. (a) A strip-chart or dot-diagram,
with a jitter to prevent points lying on top of each other; (b) a box-
and-whisker plot; (c) a histogram

pattern of the values the 915 respondents provided: these patterns can be variously termed the data distribution, **sample distribution** or empirical distribution.*

(a) The strip-chart, or dot-diagram, simply shows each data-point as a dot, but each one is given a random jitter to prevent multiple guesses of the same number lying on top of each other and obscuring the overall pattern. This clearly shows a large number of guesses in the range up to around 3,000, and then a long 'tail' of values right up to over 30,000, with a cluster at exactly 10,000.

(b) The box-and-whisker plot summarizes some essential features of the data distribution.†

(c) This histogram simply counts how many data-points lie in each of a set of intervals – it gives a very rough idea of the shape of the distribution.

These images immediately convey some distinctive features. The data distribution is highly **skewed**, meaning it is not even roughly symmetric around some central value, and has a long 'right-hand tail' due to the occurrence of some very high values. Vertical series of dots in the strip-chart also show some preference for round numbers.

But there is a problem with all these charts. The pattern of

---

* The word 'distribution' is widely used in statistics and can be ambiguous, and so I shall try to be clear what it means in each situation. Plots are implemented in the free R software.

† In this particular version of the box-and-whisker plot, the heavy central bar represents the median (the middle point), the box contains the central half of the points, while the 'whiskers' show the lowest and highest values, apart from outliers which are individually plotted.

the points means all the attention is focused on the extremely high guesses, with the bulk of the numbers being squeezed into the left-hand end. Can we present the data in a more informative way? We could throw away the extremely high values as ridiculous (and when we originally analysed this data I rather arbitrarily excluded everything above 9,000). Alternatively we could transform the data in a way that reduces the impact of these extremes, say by plotting it on what is called a **logarithmic scale,** where the space between 100 and 1,000 is the same as the space between 1,000 and 10,000.*

Figure 2.3 shows a somewhat clearer pattern, with a fairly symmetric distribution and no extreme outliers. This saves us from excluding points, which is usually not a good idea unless they are clear mistakes.

There is no 'correct' way to display sets of numbers: each of the plots we have used has some advantages: strip-charts show individual points, box-and-whisker plots are convenient for rapid visual summaries, and histograms give a good feel for the underlying shape of the data distribution.

Variables which are recorded as numbers come in different varieties:

- **Count variables**: where measurements are restricted to the integers 0, 1, 2 . . . For example, the number of

---

* To get the logarithm of a number x, we find the power of 10 that gives x, so that, for example, the logarithm of 1,000 is 3, since $10^3 = 1,000$. Logarithmic transformations are particularly appropriate when it is reasonable to assume people are making 'relative' rather than 'absolute' errors, for example because we would expect people to get the answer wrong by a relative factor, say 20% in either direction, rather than being, say, 200 beans off the true count regardless of whether they are guessing a low or high value.

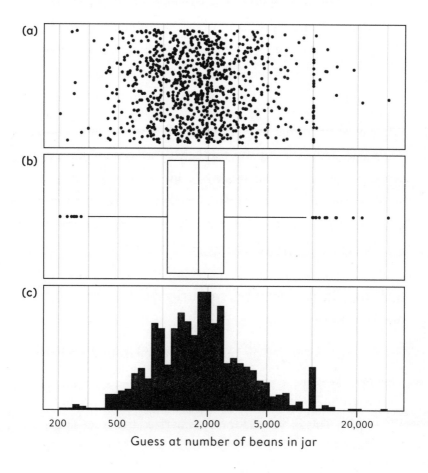

Figure 2.3
Graphical displays of the jelly-bean guesses plotted on a
logarithmic scale. (a) Strip-chart; (b) box-and-whisker plot;
(c) histogram all show a fairly symmetric pattern.

homicides each year, or guesses at the number of jelly
beans in a jar.

- **Continuous variables**: measurements that can be made,
  at least in principle, to arbitrary precision. For example,
  height and weight, each of which might vary both between
  people and from time to time. These may, of course, be
  rounded to whole numbers of centimetres or kilograms.

When a set of counts or continuous observations are reduced
to a single summary statistic, this is what we generally call their
average. We are all familiar with the idea of, for example, average wages, average exam grades and average temperatures, but
it is often unclear how to interpret these figures (particularly if
the person quoting these averages does not understand them).

There are three basic interpretations of the term 'average', sometimes jokingly referred to by the single term 'mean-
median-mode':

- **Mean**: the sum of the numbers divided by the number
  of cases.
- **Median**: the middle value when the numbers are put in
  order. This is how Galton summarized the votes of his
  crowd.*
- **Mode**: the most common value.

These are also known as measures of the location of the data
distribution.

Interpreting the term 'average' as the mean-average gives

---

* Although a 1907 correspondent to *Nature* questioned Galton's choice of the median
and claimed the mean would have given a closer estimate.

rise to the old jokes about nearly everyone having greater than the average number of legs (which is presumably around 1.99999), and people having on average one testicle. But it is not just for legs and testicles that mean-averages can be inappropriate. The mean number of reported sexual partners, and the mean income in a country, may both have little resemblance to most people's experience. This is because means are unduly influenced by a few extremely high values which drag up the total:* think Warren Beatty or Bill Gates (for sexual partners and income respectively, I should add).

Mean-averages can be highly misleading when the raw data do not form a symmetric pattern around a central value but instead are skewed towards one side like the jelly-bean guesses, typically with a large group of standard cases but with a tail of a few either very high (for example, income) or low (for example, legs) values. I can almost guarantee that, compared with people of your age and sex, you have far less than the average (mean) risk of dying next year. For example, the UK life tables report that 1% of 63-year-old men die each year before their 64th birthday, but many of those who will die are already seriously ill, and so the vast majority who are reasonably healthy will have less than this average risk.

Unfortunately, when an 'average' is reported in the media, it is often unclear whether this should be interpreted as the mean or median. For example, the UK Office for National Statistics calculates Average Weekly Earnings, which is a *mean*,

---

* Imagine three people in a room whose weekly incomes are £400, £500 and £600, so that the mean income is £1,500/3 = £500, which matches the median. Then two people who earn £5,000 a week walk in: the mean income shoots up to £11,500/5 = £2,300, while the median hardly moves, to £600.

while also reporting *median* weekly earnings by local authority. In this case it might help to distinguish between 'average income' (mean) and 'the income of the average person' (median). House prices have a very skewed distribution, with a long right-hand tail of high-end properties, which is why official house price indices are reported as medians. But these are generally reported as the 'average house price', which is a highly ambiguous term. Is this the average-house price (that is, the median)? Or the average house-price (that is, the mean)? A hyphen can make a big difference.

It is now time to reveal the results of our own wisdom of crowds experiment with the jelly beans: not as exciting as the weight of an ox, but with slightly more votes than Galton had.

Due to the data distribution having a long right-hand tail, the mean-average of 2,408 would be a poor summary, and the mode of 10,000 seems to reflect some extreme choice of round numbers. And so it is presumably better to follow Galton and use the median as a group-guess. This turns out to be 1,775 beans. The true value was . . . 1,616.[2] Just one person guessed this precisely, 45% of people guessed below 1,616, and 55% guessed above, so there was little systematic tendency for the guesses to be either on the high or low side – we say the true value lay at the 45th **percentile** of the empirical data distribution. The median, which is the 50th percentile, overestimated the true value by 1,775 – 1,616 = 159, so relative to the true answer the median was an over-estimate by around 10%, and only around 1 in 10 people got that close. So the wisdom of crowds was fairly good, getting closer to the truth than 90% of the individual people.

Describing the Spread of a Data Distribution

It is not enough to give a single summary for a distribution – we need to have an idea of the spread, sometimes known as the variability. For example, knowing the average adult male shoe size will not help a shoe firm decide the quantities of each size to make. One size does not fit all, a fact which is vividly illustrated by the seats for passengers in planes.

Table 2.1 shows a variety of summary statistics for the jelly-bean guesses, including three ways to summarize the spread. The **range** is a natural choice, but is clearly very sensitive to extreme values such as the apparently bizarre guess of 31,337 beans.* In contrast the **inter-quartile range** (IQR) is unaffected by extremes. This is the distance between the 25th and 75th percentiles of the data and so contains the 'central half' of the numbers, in this case between 1,109 and 2,599 beans: the central 'box' of the box-and-whisker plots shown above covers the inter-quartile range. Finally the **standard deviation** is a widely used measure of spread. It is the most technically complex measure, but is only really appropriate for well-behaved symmetric data† since it is also unduly influenced by outlying values. For example, removing the single (almost certainly mistaken) value of 31,337 from the data reduces the standard deviation from 2,422 to 1,398.‡

---

* This was almost certainly a mistyping of 1,337, which is a numerical rendering of the word 'leet', an internet slang term for skilled. There were nine guesses of exactly 1,337.

† The *Gini index* is a measure of spread for highly skewed data such as incomes and is widely used as a measure of inequality, but has a complex and unintuitive form.

‡ The square of the standard deviation is known as the **variance**: this is difficult to interpret directly but is mathematically useful.

Summary statistics for judgements of the number of jelly beans in a jar	Full data
Mean	2,408
Median	1,775
Mode	10,000
Range	219 to 31,337
Inter-quartile range	1,109 to 2,599
Standard deviation	2,422

Table 2.1
Summary statistics for 915 jelly-bean judgements. The true number was 1,616.

The crowd in our little experiment showed itself to have considerable wisdom, in spite of some bizarre responses. This demonstrates that data often has some errors, outliers and other strange values, but these do not necessarily need to be individually identified and excluded. It also points to the benefits of using summary measures that are not unduly affected by odd observations such as 31,337 – these are known as robust measures, and include the median and the inter-quartile range. Finally, it shows the great value of simply looking at the data, a lesson that will be reinforced by the next example.

Describing Differences between Groups of Numbers

> How many sexual partners do people in Britain report having had in their lifetime?

The purpose of this question is not simply to be nosey about people's private lives. When AIDS first became a serious concern in the 1980s, public health officials realized that there was no reliable evidence about sexual behaviour in Britain, particularly in terms of the frequency with which people changed partners, how many had multiple simultaneous partners, and what sexual practices people engaged in. This knowledge was essential to predict the spread of sexually transmitted diseases through society and to plan health services, and yet people were still quoting from the unreliable data collected by Alfred Kinsey in the US in the 1940s – who made no attempt at obtaining a representative sample.

So beginning in the late 1980s, large, careful and costly surveys of sexual behaviour were established in the UK and US, in spite of strong opposition from some quarters. In the UK, Margaret Thatcher withdrew support from a major survey of sexual lifestyles at the last minute, but those conducting the study were fortunately able to find charitable funding instead, resulting in the National Sexual Attitudes and Lifestyle Survey (Natsal) which has been carried out in the UK every ten years since 1990.

The third survey, known as Natsal-3, was carried out around 2010 and cost £7 million.[3] Table 2.2 shows the summary statistics concerning the number of (opposite-sex) sexual partners reported by people aged 35–44 in Natsal-3. It is a good exercise to use these summaries alone to try to reconstruct what the pattern of data might look like. We note that the most common single value (mode) is 1, representing those people who have only had one partner in their life, and yet there is also a massive range. This is also reflected by the substantial difference between the means and the medians, which is a telling sign of data distributions with long right-hand tails. The standard deviations are large, but this is an inappropriate measure of spread for such a data distribution, since it will be unduly influenced by a few extremely high values.

The responses of men and women may be compared by noting that men reported a mean-average of 6 more sexual partners than women, or alternatively that the average man (the median) reported 3 more sexual partners than the average woman. Or that, in relative terms, men report around 60% more partners than do women for both the mean and the median.

Reported number of sexual partners in lifetime	Men aged 35–44	Women aged 35–44
Mean	14.3	8.5
Median	8	5
Mode	1	1
Range	0 to 500	0 to 550
Inter-quartile range	4 to 18	3 to 10
Standard deviation	24.2	19.7

Table 2.2
Summary statistics for the number of (opposite-sex) sexual partners over their lifetime, as reported by 806 men and 1,215 women aged 35–44, based on interviews carried out in Natsal-3 between 2010 and 2012. Standard deviations are included for completeness, although they are inappropriate summaries of the spread of such data.

This difference might arouse our suspicions about the data. In a closed **population** with the same number of men and women with a similar age profile, it is a mathematical fact that the mean number of opposite-sex partners should be essentially the same for men and for women!* So why are men reporting so many more partners than women in this age group of 35–44? This could partly be because of men having younger partners, but also because there appears to be systematic differences in the way men and women count and report their sexual histories. We might suspect that men may be more likely to overplay their number of partners, or women underplay them, or both.

Figure 2.4 reveals the actual data distribution, which supports the impression given by the summary statistics of an extreme right-hand tail. But it is only by looking at this raw data that further important details are revealed, such as the strong tendency for both men and women to provide rounded numbers when there have been ten or more partners (except for the rather pedantic man, possibly a statistician, who said precisely, 'Forty-seven'). You may, of course, wonder about the reliability of these self-reports, and potential biases in these data are discussed in the next chapter.

Large collections of numerical data are routinely summarized and communicated using a few statistics of location and spread, and the sexual-partner example has shown that these can take us a long way in grasping an overall pattern.

---

* This is because the set of all men, and the set of all women, have the same total number of partnerships, as every partnership comprises one man and one woman. So if the groups are the same size, the mean-averages must be the same. When I discuss this with schools, I use the idea of dancing partners or handshakes.

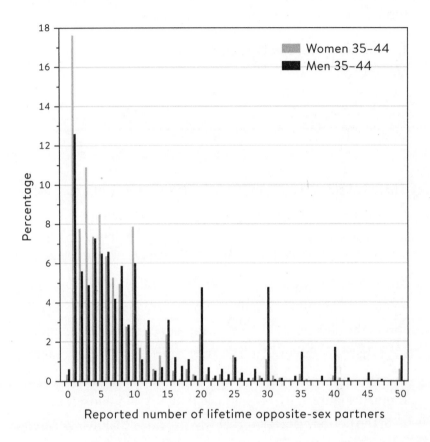

Figure 2.4
Data provided by Natsal-3 based on interviews between 2010 and 2012. The series have been truncated at 50 for reasons of space – the totals go up to 500 for both men and women. Note the clear use of round numbers for ten or more partners, and the tendency for men to report more partners than women.

However, there is no substitute for simply looking at data properly, and the next example shows that a good visualization is particularly valuable when we want to grasp the pattern in a large and complex set of numbers.

Describing Relationships Between Variables

Do busier hospitals have higher survival rates?

There is a considerable interest in the so-called 'volume effect' in surgery – the claim that busier hospitals get better survival rates, possibly since they achieve greater efficiency and have more experience. Figure 2.5 shows 30-day survival rates in UK hospitals conducting heart surgery on children plotted against the number of children being treated. Figure 2.5(a) shows the data on children aged under 1 over the period 1991–1995 that was featured at the start of the last chapter, since this age group are higher risk and were the focus of the Bristol Inquiry. Figure 2.5(b) shows the data for all children under 16 in the period 2012–2015 that was previously shown in Table 1.1 – specific data for children under 1 is not available for that period. Volume is plotted on the horizontal x-axis, and the survival rate on the vertical y-axis.*

The 1991–1995 data in Figure 2.5(a) has a clear outlier, a smaller hospital with only 71% survival. This was Bristol, whose

---

* Although the overall survival rates in the two charts are not directly comparable since the children cover different age ranges, in fact survival for children of all ages has increased from 92% to 98% over these twenty years.

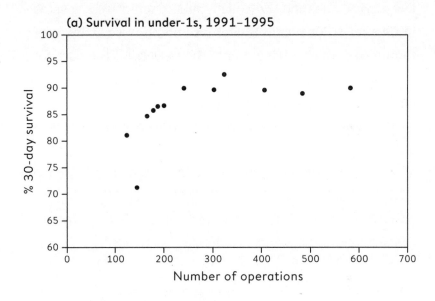

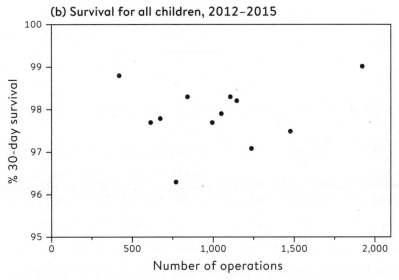

Figure 2.5
Scatter-plots of survival rates against number of operations in child heart surgery. For (a) 1991–1995, the Pearson correlation is 0.59 and the rank correlation is 0.85, for (b) 2012–2015, the Pearson correlation is 0.17 and the rank correlation is −0.03.

low survival rates and subsequent public inquiry were covered in Chapter 1. But even if Bristol is removed (try putting your thumb over the outlying point) the pattern of the data for 1991–1995 suggests that there are higher survival rates in hospitals conducting more operations.

It is convenient to use a single number to summarize a steadily increasing or decreasing relationship between the pairs of numbers shown on a scatter-plot. This is generally chosen to be the **Pearson correlation coefficient**, an idea originally proposed by Francis Galton but formally published in 1895 by Karl Pearson, one of the founders of modern statistics.*

A Pearson correlation runs between −1 and 1, and expresses how close to a straight line the dots or data-points fall. A correlation of 1 occurs if all the points lie on a straight line going upwards, while a correlation of −1 occurs if all the points lie on a straight line going downwards. A correlation near 0 can come from a random scatter of points, or any other pattern in which there is no systematic trend upwards or downwards, some examples of which are shown in Figure 2.6.

The Pearson correlation is 0.59 for the 1991–1995 data shown in Figure 2.5(a), suggesting an association of increasing volume with increasing survival. If Bristol is removed the Pearson correlation increases to 0.67, since the remaining points lie more on a straight line. An alternative measure is called **Spearman's rank correlation** after English psychologist

---

* Karl Pearson was a brilliant enthusiast for all things German: he even changed the spelling of his name from Carl to Karl, although this did not prevent him applying his statistics to ballistics in the First World War. In 1911 he founded the first Statistics Department in the world at University College London, and held the Galton Chair of Eugenics, funded from Francis Galton's will.

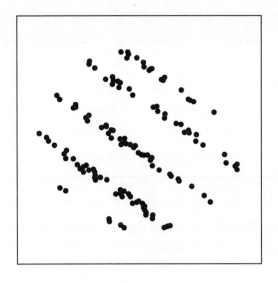

Figure 2.6
Two sets of (fictitious) data-points for which the Pearson correlation coefficients are both 0. This clearly does not mean there is no relationship between the two variables being plotted. From Alberto Cairo's wonderful Datasaurus Dozen[4].

Charles Spearman (who developed the idea of an underlying general intelligence), and depends only on the ranks of the data rather than their specific values. This means it can be near 1 or −1 if the points are close to a line that steadily increases or decreases, even if this line is not straight; the Spearman's rank correlation for the data in Figure 2.5(a) is 0.85, considerably higher than the Pearson correlation, since the points are closer to an increasing curve than a straight line.

The Pearson correlation is 0.17 for the 2012–2015 data in Figure 2.5(b), and the Spearman's rank correlation is −0.03, suggesting that there is no longer any clear relationship between the number of cases and survival rates. However, with so few hospitals the correlation coefficient can be very sensitive to individual data-points – if we remove the smallest hospital, which has a high survival rate, the Pearson correlation jumps to 0.42.

Correlation coefficients are simply summaries of association, and cannot be used to conclude that there is definitely an underlying relationship between volume and survival rates, let alone why one might exist.* In many applications the x-axis represents a quantity known as the **independent variable,** and interest focuses on its influence on the **dependent variable** plotted on the y-axis. But, as we shall explore further in Chapter 4 on causation, this presupposes the direction in which the influence might lie. Even in Figure 2.5(a)

---

* Survival rates are based on different numbers of cases and so are subject to different degrees of variability due to chance. So while a correlation can still be calculated as a description of a set of data, any formal inference needs to take into account that the data are proportions. I'll show how to do this in Chapter 6.

we cannot conclude that the higher survival rates were in any sense caused by the increased number of cases – in fact it could even be the other way round: better hospitals simply attracted more patients.

Describing Trends

What is the pattern of global population growth over the last half-century?

The population of the world is rising, and understanding the drivers of population change is of critical importance for preparing for the challenges that different countries face now and in the future. The United Nations Population Division produces estimates of population counts for all countries in the world from 1951 until now, together with projections until 2100.[5] Here we look at the worldwide trends since 1951.

Figure 2.7(a) shows simple line-graphs for the population of the world since 1951, showing around a three-fold rise to nearly 7.5 billion over this period. The rise has been driven largely from countries in Asia, but it is difficult to distinguish the patterns for other continents in Figure 2.7(a). However a logarithmic scale in Figure 2.7(b) separates out the continents, revealing the steeper gradient in Africa, and the flatter trend in other continents, particularly Europe where the population has recently been declining.

The grey lines in Figure 2.7(b) represent the changes in individual countries, but it is impossible to pick out deviations from the general upward trend. Figure 2.8 uses a simple

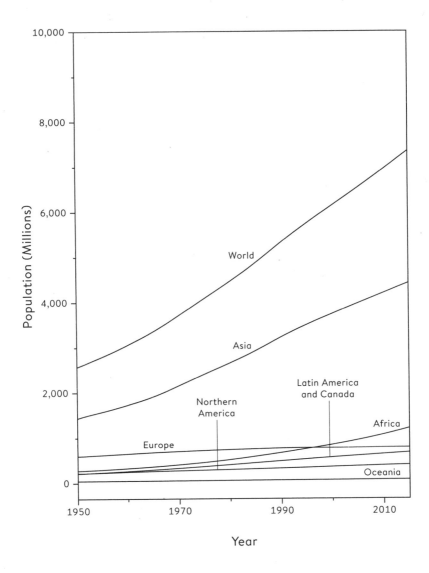

Figure 2.7
Total population for the world, continents and countries between 1950 and 2015, both sexes combined: (a) shows trends on a standard scale, (b) on a logarithmic scale, together with the trend-lines for individual countries that had a population of at least one million in 1951.

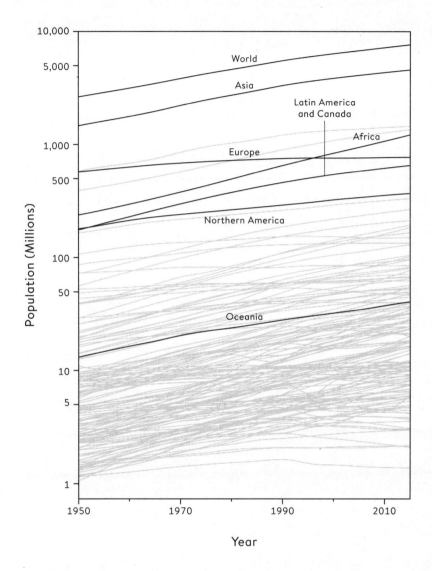

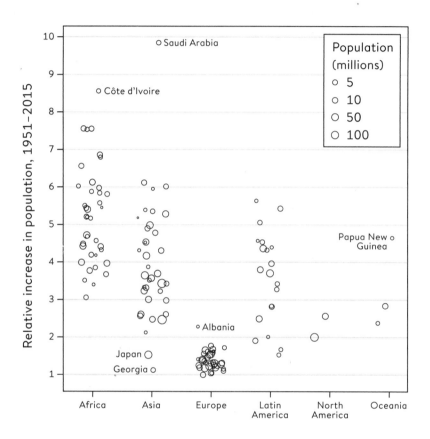

Figure 2.8
Relative increase in population between 1951 and 2015 for
countries with at least one million people in 1951.

summary of the trend in each country – the relative increase between 1951 and 2015 – where a relative increase of 4 means there are four times as many people in 2015 as there were 1951 (as happened in, for example, Liberia, Madagascar and Cameroon). Making the symbols proportional to a country's size draws the eye to the bigger countries, and grouping the countries by continents allows us to immediately detect both general clusters and outlying cases. It is always valuable to split data according to a factor – here the continents – that explains some of the overall variability.

The large increases in Africa stand out, but with wide variation and Côte d'Ivoire being an extreme case. Asia also demonstrates huge variation, reflecting the wide diversity of countries in that continent, with Japan and Georgia at one extreme and Saudi Arabia at the other, with the highest reported increase in the world. The increases in Europe have been relatively low.

Like any good graphic, this raises more questions and encourages further exploration, both in terms of identifying individual countries, and of course examining projections of future trends.

There are clearly a huge number of ways to examine such a complex data set as the UN population figures, none of which can be considered 'correct'. However, Alberto Cairo has identified four common features of a good data visualization:

1. It contains reliable information.
2. The design has been chosen so that relevant patterns become noticeable.

3. It is presented in an attractive manner, but appearance should not get in the way of honesty, clarity and depth.

4. When appropriate, it is organized in a way that enables some exploration.

The fourth feature can be facilitated by allowing the audience to interact with the visualization, and although this is difficult to illustrate in a book, the following example shows the power of personalizing a graphical display.

How popular has my name been over time?

Certain plots are so complex that it becomes difficult to spot interesting patterns with the naked eye. Take Figure 2.9, in which each line shows the rank of the popularity of a particular given name for boys born in England and Wales between 1905 and 2016.[6] This represents an extraordinary social history, and yet on its own only communicates the rapidly changing fashions in naming, with the later, denser lines suggesting a greater breadth and diversity of names since the mid-nineties.

It is only by allowing interactivity that we can pick out specific lines of personal interest. For example, I'm intrigued to see the trend for David, a name which became particularly popular in the 1920s and 1930s, possibly due to the Prince of Wales (later the short-reigned Edward VIII) being called David. But its popularity has declined precipitously – in 1953 I was one of tens of thousands of Davids, but in 2016 only 1,461 were given that name, and over 40 names were more popular.

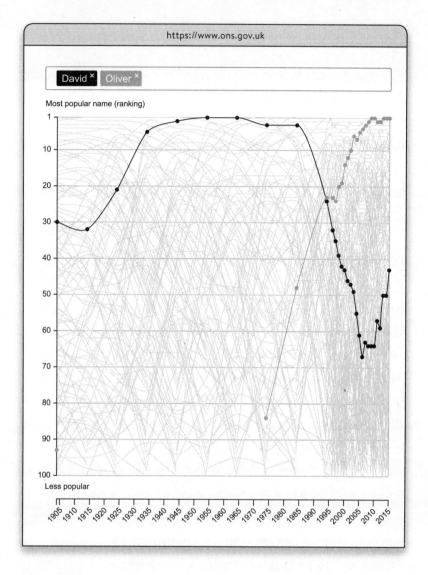

Figure 2.9

A screenshot of an interactive graph provided by the UK Office for National Statistics, showing the trend of the position of each boy's name in a league table of popularity. My rather unimaginative parents gave me the most popular boy's name in 1953, but I have since gone out of fashion, in direct contrast to Oliver. David has, however, shown some signs of recovery recently, possibly influenced by David Beckham.

Communication

This chapter has focused on summarizing and communicating data in an open and non-manipulative way; we do not want to influence our audiences' emotions and attitudes, or convince them of a certain perspective. We just want to tell it how it is, or at least how it seems to be, and while we cannot ever claim to tell the absolute truth, we can at least try to be as truthful as possible.

Of course this attempt at scientific objectivity is easier said than done. When the Statistical Society of London (later the Royal Statistical Society) was set up in 1834 by Charles Babbage, Thomas Malthus and others, they loftily declared that 'The Statistical Society will consider it to be the first and most essential rule of its conduct to exclude carefully all opinions from its transactions and publications – to confine its attention rigorously to facts – and, as far as it may be found possible, to facts which can be stated numerically and arranged in tables.'[7] From the very start they took no notice whatsoever of this stricture, and immediately starting inserting their opinions about what their data on crime, health and the economy meant and what should be done in response to it. Perhaps the best we can do now is recognize this temptation and do our best to keep our opinions to ourselves.

The first rule of communication is to shut up and listen, so that you can get to know about the audience for your communication, whether it might be politicians, professionals or the general public. We have to understand their inevitable limitations and any misunderstandings, and fight the temptation to be too sophisticated and clever, or put in too much detail.

The second rule of communication is to know what you want to achieve. Hopefully the aim is to encourage open debate, and informed decision-making. But there seems no harm in repeating yet again that numbers do not speak for themselves; the context, language and graphic design all contribute to the way the communication is received. We have to acknowledge we are telling a story, and it is inevitable that people will make comparisons and judgements, no matter how much we only want to inform and not persuade. All we can do is try to pre-empt inappropriate gut reactions by design or warning.

Storytelling with Statistics

This chapter has introduced the concept of data visualization, sometimes known as dataviz. These techniques are often used for researchers, or for fairly sophisticated audiences, using a standard armoury of plots that are selected for their value in gaining understanding and for exploring the data, rather than their purely visual appeal. When we have worked out the important messages in the data that we want to communicate, we might then go on to use infographics, or infoviz, to grab the attention of the audience and tell a good story.

Sophisticated infographics regularly appear in the media, but Figure 2.10 shows a fairly basic example which tells a strong story of social trends by bringing together the responses to three questions in the UK's National Survey of Sexual Attitudes and Lifestyles (Natsal-3) in 2010; at what age did women and men first have sex, first start co-habiting, and have their first child?[8] The median ages for each of these life events are plotted against the women's year of birth, and the three points connected with a heavy vertical line. The steady lengthening of this line between

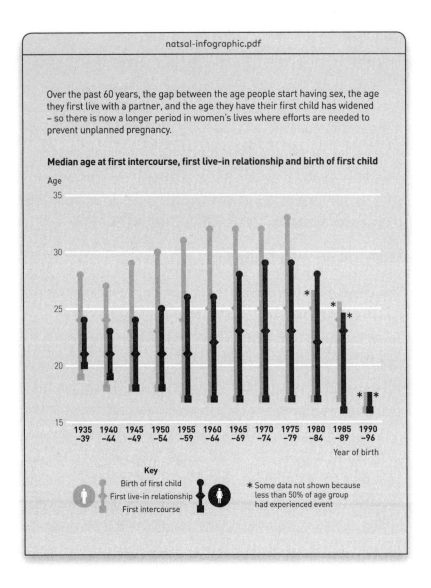

natsal-infographic.pdf

Over the past 60 years, the gap between the age people start having sex, the age they first live with a partner, and the age they have their first child has widened – so there is now a longer period in women's lives where efforts are needed to prevent unplanned pregnancy.

Median age at first intercourse, first live-in relationship and birth of first child

Figure 2.10

Infographic based on data from the third UK National Survey of Sexual Attitudes and Lifestyles (Natsal-3) – the lesson from the data is pointed out both visually and verbally.

women born in the 1930s and those in the 1970s displays the increased period in which effective contraception is necessary.

Even more advanced are dynamic graphics, in which movement can be used to reveal patterns in the changes over time. The master of this technique was Hans Rosling, whose TED talks and videos set a new standard of storytelling with statistics, for example by showing the relationship between changing wealth and health through the animated movement of bubbles representing each country's progress from 1800 to the present day. Rosling used his graphics to try to correct misconceptions about the distinction between 'developed' and 'undeveloped' countries, with the dynamic plots revealing that, over time, almost all countries moved steadily along a common path towards greater health and prosperity.*[9]

This chapter has demonstrated a continuum from simple descriptions and plots of raw data, through to complex examples of storytelling with statistics. Modern computing means that data-visualization is becoming easier and more flexible; and since summary statistics can hide as well as illuminate, appropriate graphical displays are essential. Nevertheless, summarizing and communicating the raw numbers is only the first stage in the process of learning from data. To get further along this path, we need to address the fundamental idea of what we are trying to achieve in the first place.

---

* Unfortunately a static book in grayscale is an unsuitable medium for displaying his work, so I can only encourage you to check gapminder.org. Rosling was once arguing on TV with a Danish journalist who was parroting the sort of misconceptions about the world that Hans spent his life trying to counter. Hans simply replied, 'These facts are not up for discussion. I am right, and you are wrong' – which, for statistics, is unusually straight speaking.

Summary

- A variety of statistics can be used to summarize the empirical distribution of data-points, including measures of location and spread.
- Skewed data distributions are common, and some summary statistics are very sensitive to outlying values.
- Data summaries always hide some detail, and care is required so that important information is not lost.
- Single sets of numbers can be visualized in strip-charts, box-and-whisker plots and histograms.
- Consider transformations to better reveal patterns, and use the eye to detect patterns, outliers, similarities and clusters.
- Look at pairs of numbers as scatter-plots, and time-series as line-graphs.
- When exploring data, a primary aim is to find factors that explain the overall variation.
- Graphics can be both interactive and animated.
- Infographics highlight interesting features and can guide the viewer through a story, but should be used with awareness of their purpose and their impact.

CHAPTER 3
Why Are We Looking at Data Anyway?
Populations and Measurement

> How many sexual partners have people in Britain
> *really* had?

The last chapter showed some remarkable results from a recent UK survey in which people reported the number of sexual partners they had had in their lifetime. Plotting these responses revealed various features, including a (very) long tail, a tendency to use round numbers such as 10 and 20, and more partners reported by men than women. But the researchers who spent millions of pounds collecting this data were not really interested in what these particular respondents said – after all, they were guaranteed complete anonymity. Their responses were a means to an end, which was to say something about the overall pattern of sexual partnerships in Britain – those of the millions of people who were *not* questioned about their sexual behaviour.

It is no trivial matter to go from the actual responses collected in a survey to conclusions about the whole of Britain. Actually, this is incorrect – it is incredibly easy to just claim that what these respondents say accurately represents what is really going on in the country. Media surveys about sex, where people volunteer to fill in forms on websites about what they say they get up to behind closed doors, do this all the time.

The process of going from the raw responses in the survey to claims about the behaviour of the whole country can be broken down into a series of stages:

1. The recorded *raw data* on the number of sexual partners that our survey participants report tells us something about . . .
2. The *true number* of partners of people in our *sample*, which tells us something about . . .
3. The number of partners of people in the *study population* – the ones who could potentially have been included in our survey – which tells us something about . . .
4. The number of sexual partners for people in Britain, which is our *target population*.

Where are the weakest points in this chain of reasoning? Going from the raw data (Stage 1) to the truth about our sample (Stage 2) means making some strong assumptions about how accurate respondents are when they say how many partners they have had, and there are many reasons to doubt them. We have already seen an apparent tendency for men to overstate, and women to understate, their partner count, possibly due to women not including partnerships they would rather forget, different tendencies to round up or round down, poor memory, and simple 'social acceptability bias'.*

---

* Some evidence for such a bias was obtained in a randomized experiment on US students, where women who were wired to a lie detector tended to admit to having had more partners than those guaranteed anonymity, whereas the effect was not found in men. The participants were not told that the lie detector was fake.

Going from our sample (Stage 2) to the study population (Stage 3) is perhaps the most challenging step. We first need to be confident that the people asked to take part in the survey are a random sample from those who are eligible: this should be fine for a well-organized study like Natsal. But we also need to assume that the people who actually agree to take part are representative, and this is less straightforward. The surveys have around a 66% response rate, which is remarkably good given the nature of the questions. However there is some evidence that participation rates are slightly lower in those who are not so sexually active, possibly counterbalanced by the difficulty in getting interviews with more unconventional members of society.

Finally, going from the study population (Stage 3) to the target population (Stage 4) is more straightforward, provided we can assume that the people who could potentially have been asked to participate represent the adult population of Britain. In Natsal's case this should be assured by their careful experimental design, based on a random sample of households, although this does mean that people in institutions such as prisons, the services or nunneries were not included.

By the time we have worked through all the things that can go wrong, it might be enough to make anyone sceptical about making any general claims about the true sexual behaviour of the country, based on what we are told by the respondents to the survey. But the whole point of statistical science is to smooth progress through these stages and finally, with due humility, be able to say what we can and cannot learn from data.

Learning from Data – the Process of 'Inductive Inference'

The preceding chapters have assumed you have a problem, you get some data, you look at it, and then summarize it concisely. Sometimes the counting, measuring and describing is an end in itself. For instance, if we just want to know how many people passed through the Accident & Emergency department last year, the data can tell us the answer.

But often the question goes beyond simple description of data: we want to learn something bigger than just the observations in front of us, whether it is to make predictions (how many will come next year?), or say something more basic (why are the numbers increasing?).

Once we want to start generalizing from the data – learning something about the world outside our immediate observations – we need to ask ourselves the question, 'Learn about what?' And this requires us to confront the challenging idea of **inductive inference.**

Many people have a vague idea of *deduction*, thanks to Sherlock Holmes using deductive reasoning when he coolly announces that a suspect must have committed a crime. In real life deduction is the process of using the rules of cold logic to work from general premises to particular conclusions. If the law of the country is that cars should drive on the right, then we can deduce that on any particular occasion it is best to drive on the right. But *induction* works the other way, in taking particular instances and trying to work out general conclusions. For example, suppose we don't know the customs in a community about kissing female friends on the cheek, and we have to try to work it out by observing whether people kiss once, twice, three times, or not at all. The crucial distinction is that deduction is logically certain, whereas induction is generally uncertain.

Figure 3.1 represents inductive inference as a generic diagram, showing the steps involved in going from data to the eventual target of our investigation: as we have seen, the data collected in the sex survey tells us about the behaviour of our sample, which we use to learn about the people who could have been recruited to the survey, from which we make some tentative conclusions about sexual behaviour in the whole country.

Of course, it would be ideal if we could go straight from looking at the raw data to making general claims about the target population. In standard statistics courses, observations are assumed to be drawn perfectly randomly and directly from the population of direct interest. But this is rarely the case in real life, and therefore we need to consider the entire process of going from raw data to our eventual target. And, as we have seen with the sex survey, problems can occur at each of the different stages.

Going from data (Stage 1) to the sample (Stage 2): these are problems of *measurement*: is what we record in our data an accurate reflection of what we are interested in? We want our data to be:

- Reliable, in the sense of having low variability from occasion to occasion, and so being a precise or repeatable number.

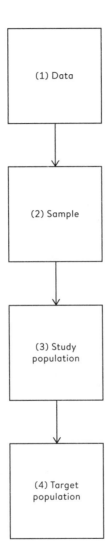

Figure 3.1
Process of inductive inference: each arrow can be interpreted as
'tells us something about'[1]

- Valid, in the sense of measuring what you really want to measure, and not having a systematic bias.

For example, the adequacy of the sex survey depends on people giving the same or very similar answers to the same question each time they are asked, and this should not depend on the style of the interviewer or the vagaries of the respondent's mood or memory. This can be tested to some extent by asking specific questions both at the start and end of the interview. The quality of the survey also requires the interviewees to be honest when they report their sexual activity, and not either systematically exaggerate or downplay their experiences. All these are fairly strong demands.

A survey would not be valid if the questions were biased in favour of a particular response. For example, in 2017 budget airline Ryanair announced that 92% of their passengers were satisfied with their flight experience. It turned out that their satisfaction survey only permitted the answers, 'Excellent, very good, good, fair, OK'.*

We have seen how positive or negative framing of numbers can influence the impression given, and similarly the framing of a question can influence the response. For example, a UK survey in 2015 asked people whether they supported or opposed 'giving 16- and 17-year-olds the right to vote' in the

---

* After someone from the Royal Statistical Society criticized their survey methods, a spokesman for Ryanair's boss Michael O'Leary said, 'Ninety-five per cent of Ryanair customers haven't heard of the Royal Statistical Society, 97 per cent don't care what they say and 100 per cent said it sounds like their people need to book a low-fare Ryanair holiday.' In another contemporary survey, Ryanair was voted bottom out of twenty European airlines (but this had its own reliability problems, being held at a time when Ryanair had cancelled large numbers of flights).

referendum on whether to leave the European Union, and 52% supported the idea while 41% opposed it. So the majority favoured this proposal when framed in terms of recognizing rights and empowering younger people.

But when the same respondents were asked the (logically identical) question of whether they supported or opposed 'reducing the voting age from 18 to 16' for the referendum, the proportion supporting the proposal dropped to 37%, with 56% opposing. So when framed in terms of a more risky liberalization, the proposal was opposed by the majority, a reversal in opinion brought about by simple rewording of the question.[2]

The responses to questions can also be influenced by what has been asked beforehand, a process known as priming. Official surveys of wellbeing estimate that around 10% of young people in the UK consider themselves lonely, but an online questionnaire by the BBC found the far higher proportion of 42% among those choosing to answer. This figure may have been inflated by two factors: the self-report nature of the voluntary 'survey', and the fact that the question about loneliness had been preceded by a long series of enquiries as to whether the respondent in general felt a lack of companionship, isolated, left out, and so on, all of which might have primed them to give a positive response to the crucial question of feeling lonely.[3]

Going from sample (Stage 2) to study population (Stage 3): this depends on the fundamental quality of the study, also known as its **internal validity**: does the sample we observe accurately reflect what is going on in the group we are actually studying? This is where we come to the crucial way of avoiding bias: random sampling. Even children understand what it means to pick something at random: closing your eyes and reaching into a jumbled bag of sweets and seeing which colour comes out, or pulling a number out of a hat to see who gets a prize or a treat (or doesn't). It has been used for millennia as a way of ensuring fairness and justice, when it is known as sortition,* and has been used as a way of allocating rewards,† running lotteries, and appointing people with power such as officials and jurors. It has also been involved in more sobering duties, such as choosing which young people should go off to war, or who to eat in a lifeboat lost at sea.

George Gallup, who essentially invented the idea of the opinion poll in the 1930s, came up with a fine analogy for the value of random sampling. He said that if you have cooked a large pan of soup, you do not need to eat it all to find out if it needs more seasoning. You can just taste a spoonful, *provided you have given it a good stir*. A literal proof of this idea was provided by the 1969 Vietnam War draft lottery, which had to provide an ordered list of birthdays, and then men whose birthday was at the top of the list would be drafted first to go to Vietnam, and so on down the list. In a public attempt to make the process fair, 366 capsules were prepared, each containing a unique birthday, and capsules were intended to be

---

* Not to be confused with *sortilege*, which is a form of divination in which apparently chance phenomena are used to determine divine will or future fortune – this is also known as *cleromancy*. Examples exist in many cultures, including fortune-telling using tea-leaves or chicken entrails, biblical casting of lots to determine the will of God, and divination using the *I Ching*.

† 'Then said Jesus, Father, forgive them; for they know not what they do. And they parted his clothing, and cast lots', Luke 23:34.

picked from a box at random. But the capsules were put in the box in order of the month of the birthday, and were not properly mixed up. This might not have caused a problem if the men drawing out the capsules had delved down into the box, but as a remarkable video shows, they tended to take them from the top.[4] The result was that it was bad luck to be born later in the year: 26 out of 31 birthdays in December ended up drafted, compared to only 14 in January.

The idea of adequate 'stirring' is crucial: if you want to be able to generalize from the sample to the population, you need to make sure your sample is representative. Just having masses of data does not necessarily help guarantee a good sample and can even give false reassurance. For example, polling companies performed miserably in the 2015 UK general election, even though they had sampled thousands of potential voters. A later inquiry blamed non-representative sampling, particularly from telephone polls: not only did landlines make up the majority of numbers called, but less than 10% of those who were called actually responded. This is hardly likely to be a representative sample.

Going from study population (Stage 3) to target population (Stage 4): finally, even with perfect measurement and a meticulous random sample, the results may still not reflect what we wanted to investigate in the first place if we have not been able to ask the people in whom we are particularly interested. We want our study to have **external validity**.

An extreme example is when our target population comprises people, whereas we have only been able to study animals, such as the effect of a chemical on mice. Less dramatic is when clinical trials of new drugs have been conducted only on adult men, but the drug is then used 'off-label' on women and children. We would like to know the effects on everyone, but this cannot be solved by statistical analysis alone – we inevitably need to make assumptions and be very cautious.

When We Have All the Data

Although the ideas of learning from data are neatly illustrated by looking at surveys, in fact much of the data used today is not based on random sampling, or in fact any sampling at all. Routinely collected data on, say, online purchasing or social transactions, or for administering a system such as education or policing, can be re-purposed to help us understand about what is going on in the world. In these situations we have all the data. In terms of the process of induction shown in Figure 3.1, there is no gap between Stages 2 and 3 – the 'sample' and the study population are essentially the same. This does avoid any concern about having a small sample size, but many other problems can still remain.

Consider the question of how much crime there is in Britain, and the politically sensitive issue of whether it is going up or going down. There are two main sources of data – one survey-based and one administrative. First, the Crime Survey for England and Wales is a classic piece of sampling in which around 38,000 people are questioned each year about their experiences of crime. Just like the Natsal sex survey, problems can arise when using the actual reports (Stage 1) to draw conclusions about their true experiences (Stage 2), since respondents may not tell the truth – say about drug crime in which they have themselves participated. Then we need to assume the sample is representative of the eligible population and take into account its limited size (Stage 2 to Stage 3), and finally acknowledge that the study design is not reaching some part of the overall target population, such as the fact that nobody under 16 or living in a communal residence is questioned (Stage 3 to Stage 4). Nevertheless, with suitable caveats, the Crime Survey for England and Wales is a 'designated national statistic' and used for monitoring long-term trends.[5]

The second source of data comprises the reports of crimes recorded by the police. This is done for administrative purposes and is not a sample: since every crime that is recorded in the country can be counted, the 'study population' is the same as the sample. Of course we still have to assume that the data recorded truly represent what happened to those victims who report crimes (Stage 1 to Stage 2), but the major problem occurs when we want to claim that the data on the study population – people who reported crimes – represents the target population of all crimes committed in England and Wales. Unfortunately, police-recorded crime systematically misses cases which the police do not record as a crime or which have not been reported by the victim; illegal drug use, for example, and people who choose not to report thefts and vandalism in case their area suffers a decline in property values. As an extreme example, after a report in November 2014 criticized the police's recording practices, the number of recorded sexual offences rose from 64,000 in 2014 to 121,000 in 2017: a near doubling in three years.

It is hardly surprising that these two different sources of data can come up with rather different conclusions about trends: for example, the Crime Survey estimated that crime fell by 9% between 2016 and 2017, while the police recorded

13% more offences. Which should we believe? Statisticians have more confidence in the survey, and concerns about the reliability of police-recorded crime data led it to lose its designation as a national statistic in 2014.

When we have all the data, it is straightforward to produce statistics that describe what has been measured. But when we want to use the data to draw broader conclusions about what is going on around us, then the quality of the data becomes paramount, and we need to be alert to the kind of systematic biases that can jeopardize the reliability of any claims.

Whole websites are dedicated to listing the possible biases that can occur in statistical science, from allocation bias (systematic differences in who gets each of two medical treatments being compared) to volunteer bias (people volunteering for studies being systematically different from the general population). Many of these are fairly common sense, although in Chapter 12 we shall see some more subtle ways in which statistics can be done badly. But first we should consider ways of describing our ultimate aim – the target population.

The 'Bell-Shaped Curve'

A friend in the US has just given birth to a full-term baby weighing 6 lb 7 oz (2.91 kg). She has been told this is below average, and is concerned. Is the weight unusually low?

We have already discussed the concept of a data distribution – the pattern the data makes, sometimes known as the empirical or sample distribution. Next we must tackle the concept of a **population distribution** – the pattern in the whole group of interest.

Consider an American woman who has just given birth. We might think of her baby as having been drawn, as a sort of sample of only one person, from the entire population of babies recently born to non-Hispanic white women in the US (her race is important, since birth weights are reported for different races). The population distribution is the pattern made by the birth weights of all these babies, which we can obtain from the US National Vital Statistics System's report on the weights of over a million babies born at full-term in the US in 2013 to non-Hispanic white women – although this is not the entire set of contemporary births, it is such a large sample that we can take it as the population.[6] These birth weights are only reported as the numbers in groups spanning 500 g, and are shown in Figure 3.2(a).

The weight of your friend's baby is indicated as a line at 2,190 grams, and its position in the distribution can be used to assess whether its weight is 'unusual'. The shape of this distribution is important. Measurements such as weight, income, height, and so on can, at least in principle, be as fine-grained as desired, and so can be considered 'continuous' quantities whose population distributions are smooth. The classic example is the 'bell-shaped curve', or **normal distribution**, first explored in detail by Carl Friedrich Gauss in 1809 in the context of measurement errors in astronomy and surveying.*

---

* Gauss's derivation was not based on empirical observation, but was a theoretical form of measurement error that would justify his statistical methods.

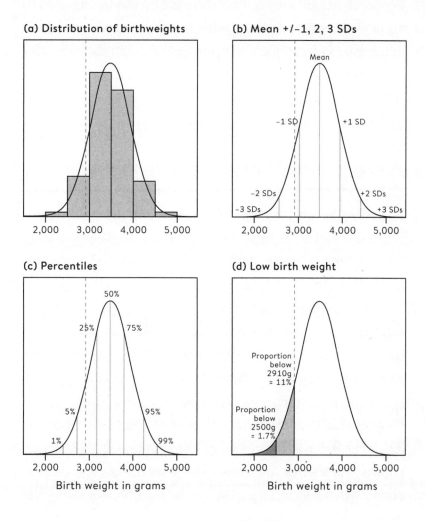

(a) Distribution of birthweights

2,000   3,000   4,000   5,000

(b) Mean +/-1, 2, 3 SDs

Mean

-1 SD   +1 SD

-2 SDs   +2 SDs
-3 SDs   +3 SDs

2,000   3,000   4,000   5,000

(c) Percentiles

50%

25%   75%

5%   95%

1%   99%

2,000   3,000   4,000   5,000

Birth weight in grams

(d) Low birth weight

Proportion
below
2910g
= 11%

Proportion
below
2500g
= 1.7%

2,000   3,000   4,000   5,000

Birth weight in grams

Figure 3.2
(a) The distribution of birth weights of 1,096,277 children of non-Hispanic white women in the US in 2013, born at 39–40 weeks' gestation, with a normal curve with the same mean and standard deviation as the recording weights in the population. A baby weighing 2,910g is shown as the dashed line. (b) The mean ± 1, 2, 3 standard deviations (SDs) for the normal curve. (c) Percentiles of the normal curve. (d) The proportion of low-birth-weight babies (dark shaded area), and babies less than 2,910g (light shaded area).

Theory shows that the normal distribution can be expected to occur for phenomena that are driven by large numbers of small influences, for example a complex physical trait that is not influenced by just a few genes. Birth weight, when looked at for a single ethnic group and gestation period, might be considered such a trait, and Figure 3.2(a) shows a normal curve with the same mean and standard deviation as the recorded weights. The smooth normal curve and the histogram are gratifyingly close, and other complex traits such as height and cognitive skills also have approximately normal population distributions. Other, less natural phenomena may have population distributions that are distinctly non-normal and often feature a long right-hand-tail, income being a classic example.

The normal distribution is characterized by its **mean**, or **expectation**, and its standard deviation, which as we have seen is a measure of spread – the best-fitting curve in Figure 3.2(a) has a mean of 3,480 g (7 lb 11 oz) and a standard deviation of 462 g (1 lb). We see that the measures used to summarize data sets in Chapter 2 can be applied as descriptions of a population too – the difference is that terms such as mean and standard deviation are known as **statistics** when describing a set of data, and **parameters** when describing a population. It is an impressive achievement to be able to summarize over 1,000,000 measurements (that is, over a million births) by just these two quantities.

A great advantage of assuming a normal form for a distribution is that many important quantities can be simply obtained from tables or software. For example, Figure 3.2(b) shows the position of the mean and 1, 2 and 3 standard deviations each side of the mean. From the mathematical properties of the normal distribution, we know that roughly 95% of the population will be contained in the interval given by the mean ± two standard deviations, and 99.8% in the central ± three standard deviations. Your friend's baby is around 1.2 standard deviations below the mean – this is also known as her **Z-score**, which simply measures how many standard deviations a data-point is from the mean.

The mean and standard deviation can be used as summary descriptions for (most) other distributions, but other measures may also be useful. Figure 3.2 (c) shows selected **percentiles** calculated from the normal curve: for example the 50th percentile is the median, the point which splits the population in half and which could be said to be the weight of an 'average' baby – this is the same as the mean in the case of a symmetric distribution such as the normal curve. The 25th percentile (3,167 g) is the weight under which 25% of babies lie – the 25th and 75th percentiles (3,791 g) are known as the **quartiles**, and the distance between them (624 g), known as the interquartile range, is a measure of the spread of the distribution. Again, these are exactly the same summaries as used in Chapter 2, but here applied to populations rather than samples.

Your friend's baby lies on the 11th percentile, which means that 11% of full-term babies born to non-Hispanic white women will weigh less – Figure 3.2(d) shows this 11% as a light grey shaded area. Birth-weight percentiles are of practical importance, since your friend's baby's weight will be monitored relative to the growth expected for babies on the 11th percentile,* and a drop in the baby's percentile may be a cause for concern.

---

* Although the distributions used for this monitoring will be somewhat more sophisticated than the normal.

For medical rather than statistical reasons, babies below 2,500 g are considered 'low birth weight', and those below 1,500 g 'very low birth weight'. Figure 3.2(d) shows that we would expect 1.7% of babies in this group to be low birth weight – in fact the actual number was 14,170 (1.3%), in close agreement with the prediction from the normal curve. We note that this particular group of full-term births to non-Hispanic white mothers has a very small rate of low birth weights – the overall rate for all births in the US in 2013 was 8%, while the rate in black women was 13%, a notable difference between races.

Perhaps the most crucial lesson from this example is that the dark-grey shaded area in Figure 3.2(d) plays two roles:

1. It represents the *proportion* of this population of babies being low birth weight.
2. It is also the *probability* that a randomly chosen baby in 2013 weighs less than 2,500 g.

So a population can be thought of as a physical group of individuals, but also as providing the **probability distribution** for a random observation. This dual interpretation will be fundamental when we come to more formal statistical inference.

Of course in this case we know the shape and parameters of the population, and so we can say something both about the proportions in the population, and the chances of different events occurring for a random observation. But the whole point of this chapter is that we do not generally know about populations, and so want to follow the inductive process and go the other way around, from data to population. We have seen that the standard measures of mean, median, mode, and so on, which we developed for samples, extend to whole populations – but the difference is that we do not know what they are. And that is the challenge we face in the next chapter.

What Is the Population?

The stages of induction outlined above work well with planned surveys, but a lot of statistical analysis does not fit as easily into this framework. We have seen that, particularly when using administrative records such as police reports on crime, we may have all the possible data. But although there is no sampling, the idea of an underlying population can still be valuable.

Consider the children's heart surgery data in Chapter 1. We made the rather bold assumption there were no measurement problems – in other words we have a complete collection of both the operations and 30-day survivors in each hospital. So our knowledge of the sample (Stage 2) is perfect.

But what is the study population? We have data on all the children and all the hospitals, and so there is no larger group from which they have been sampled. Although the idea of a population is usually introduced rather casually into statistics courses, this example shows it is a tricky and sophisticated idea that is worth exploring in some detail, as a lot of important ideas build on this concept.

There are three types of populations from which a sample might be drawn, whether the data come from people, transactions, trees, or anything else.

- A *literal* population. This is an identifiable group, such as when we pick a person at random when polling. Or there may be a group of individuals who could be measured, and although we don't actually pick one at random, we have data from volunteers. For example, we might consider the people who guessed at the number of jelly beans as a sample from the population of all maths nerds who watch YouTube videos.

- A *virtual* population. We frequently take measurements using a device, such as taking someone's blood pressure or measuring air pollution. We know we could always take more measurements and get a slightly different answer, as you will know if you have ever taken repeat blood pressure measurements. The closeness of the multiple readings depends on the precision of the device and the stability of the circumstances – we might think of this as drawing observations from a virtual population of all the measurements that could be taken if we had enough time.

- A *metaphorical* population, when there is no larger population at all. This is an unusual concept. Here we act *as if* the data-point were drawn from some population at random, but it clearly is not – as with the children having heart surgery: we did not do any sampling, we have all the data, and there is no more we could collect. Think of the number of murders that occur each year, the examination results for a particular class, or data on all the countries of the world – none of these can be considered as a sample from an actual population.

The idea of a metaphorical population is challenging, and it may be best to think of what we have observed as having been drawn from some imaginary space of possibilities. For example, the history of the world is what it is, but we can imagine history having played out differently, and we happen to have ended up in just one of these possible states of the world. This set of all the alternative histories can be considered a metaphorical population. To be more concrete, when we looked at childhood heart surgery in the UK between 2012 and 2015, we had all the data on surgery in those years and knew how many deaths and how many survivors there were. Yet we can imagine counterfactual histories in which different individuals might have survived, through unforeseeable circumstances that we tend to call 'chance'.

It should be apparent that rather few applications of statistical science actually involve literal random sampling, and that it is increasingly common to have all the data that is potentially available. Nevertheless it is extremely valuable to keep hold of the idea of an imaginary population from which our 'sample' is drawn, as then we can use all the mathematical techniques that have been developed for sampling from real populations.

Personally, I rather like acting as if all that occurs around us is the result of some random pick from the all the possible things that could happen. It is up to us whether we choose to believe it is truly chance, whether it the will of a god or gods, or any other theory of causation: it makes no difference to the mathematics. This is just one of the mind-stretching requirements for learning from data.

Summary

- Inductive inference requires working from our data, through study sample and study population, to a target population.
- Problems and biases can crop up at each stage of this path.
- The best way to proceed from sample to study population is to have drawn a random sample.
- A population can be thought of as a group of individuals, but also as providing the probability distribution for a random observation drawn from that population.
- Populations can be summarized using parameters that mirror the summary statistics of sample data.
- Often data does not arise as a sample from a literal population. When we have all the data there is, then we can imagine it drawn from a metaphorical population of events that could have occurred, but didn't.

CHAPTER 4

What Causes What?

> Does going to university increase the risk of getting
> a brain tumour?

Epidemiology is the study of how and why diseases occur in
the population, and Scandinavian countries are an epidemi-
ologist's dream. This is because everyone in those countries
has a personal identity number which is used when register-
ing for health care, education, tax, and so on, and this allows
researchers to link all these different aspects of people's lives
together in a way that would be impossible (and perhaps politically controversial) in other countries.

A typically ambitious study was conducted on over 4 million
Swedish men and women whose tax and health records were
linked over eighteen years, which enabled the researchers to
report that men with a higher socioeconomic position had a
slightly increased rate of being diagnosed with a brain tumour.
This was one of those worthy but rather unexciting studies
that would typically not attract much attention, so a university
communications officer thought it would be more interesting
to say in a press release that 'High levels of education are linked
to heightened brain tumour risk', even though the study was

about socioeconomic position rather than education. And by the time this got to the general public, a subeditor in a newspaper produced the classic headline, 'Why Going to University Increases Risk of Getting a Brain Tumour'.[1]

For anyone who has spent time accumulating academic qualifications, this newspaper headline could have been alarming. But should we be concerned? This is a huge study based on a registry of the complete eligible population – not a sample – so we can confidently conclude that slightly more brain tumours really were found in more-educated people. But did all that sweating in the library overheat the brain and lead to some strange cell mutations? In spite of the newspaper headline, I doubt it. And to give them credit, the authors of the paper doubted it too, adding, 'Completeness of cancer registration and detection bias are potential explanations for the findings.' In other words, wealthy people with higher education are more likely to be diagnosed and get their tumour registered, an example of what is known as **ascertainment bias** in epidemiology.

'Correlation Does Not Imply Causation'

We saw in the last chapter how Pearson's correlation coefficient measures how close the points on a scatter-plot are to a straight line. When considering English hospitals conducting children's heart surgery in the 1990s, and plotting the number of cases against their survival, the high correlation showed that bigger hospitals were *associated* with lower mortality. But we could not conclude that bigger hospitals *caused* the lower mortality.

This cautious attitude has a long pedigree. When Karl Pearson's newly developed correlation coefficient was being discussed in the journal *Nature* in 1900, a commentator warned that 'correlation does not imply causation'. In the succeeding century this phrase has been a mantra repeatedly uttered by statisticians when confronted by claims based on simply observing that two things tend to vary together. There is even a website that automatically generates idiotic associations, such as the delightful correlation of 0.96 between the annual per-capita consumption of mozzarella cheese in the US between 2000 and 2009, and the number of civil engineering doctorates awarded in each of those years.[2]

There seems to be a deep human need to explain things that happen in terms of simple cause–effect relationships – I am sure we could all construct a good story about all those new engineers gorging on pizzas. There is even a word for the tendency to construct reasons for a connection between what are actually unrelated events – *apophenia* – with the most extreme case being when simple misfortune or bad luck is blamed on others' ill-will or even witchcraft.

Unfortunately, or perhaps fortunately, the world is a bit more complicated than simple witchcraft. And the first complication comes in trying to work out what we mean by 'cause'.

What Is 'Causation' Anyway?

Causation is a deeply contested subject, which is perhaps surprising as it seems rather simple in real life: we do something, and that leads to something else. I jammed my thumb in the car door, and now it hurts.

But how do we know that my thumb would not have hurt anyway? Perhaps we can think of what is known as a **counterfactual**. If I hadn't jammed my thumb in the door, then my thumb would not hurt. But this will always be an assumption, requiring the rewriting of history, since we can never really know for certain what I might have felt (although in this case I might be fairly confident that my thumb would not suddenly start hurting of its own accord).

This gets even trickier when we allow for the unavoidable variability that underlies everything interesting in real life. For example, the medical community now agrees that smoking cigarettes causes lung cancer, but it took decades for doctors to come to this conclusion. Why did it take so long? Because most people who smoke do not get lung cancer. And some people who do not smoke do get lung cancer. All we can say is that you are more likely to get lung cancer if you smoke than if you do not smoke, which is one reason why it took so long for laws to be enacted to restrict smoking.

So our 'statistical' idea of causation is not strictly deterministic. When we say that X causes Y, we do not mean that every time X occurs, then Y will too. Or that Y will only occur if X occurs. We simply mean that if we intervene and force X to occur, then Y tends to happen more often. So we can never say that X caused Y in a specific case, only that X increases the proportion of times that Y happens. This has two vital consequences for what we have to do if we want to know what causes what. First, in order to infer causation with real confidence, we ideally need to intervene and perform experiments. Second, since this is a statistical or stochastic world, we need to intervene more than once in order to amass evidence.

And that leads us naturally to a delicate topic: conducting medical experiments on large numbers of people. Few of us might relish the idea of being experimented on, especially when life and death are concerned. Which makes it all the more remarkable that thousands of people have been willing to be part of huge studies in which neither they nor their doctor knew which treatment they would end up getting.

Do statins reduce heart attacks and strokes?

Every day I take a little white pill – a statin – because I have been told it lowers cholesterol and so reduces the risk of heart attacks and strokes. But what is its effect on me personally? I am almost certain that it causes my low-density cholesterol (LDL) to drop, since I was told it reduced soon after I started taking the tablets. This drop in LDL is a direct, essentially deterministic effect that I can assume is caused by the statin.

But I will never know if this daily ritual does me any good in the long run; it depends on which of my many possible future lives actually occurs. If I never have a heart attack or a stroke, I will have no idea whether I would have never had one even if I had not taken the tablets, and all this pill-popping for years was a waste of time. If I do have a heart attack or a stroke, I will not know if this event was delayed by taking the statin. All I can ever know is that, on average, it benefits a large group of people like me, and this knowledge is based on large clinical trials.

The purpose of a clinical trial is to carry out a 'fair test' that properly determines causation and estimates the average effect of a new medical treatment, without introducing biases that could give us the wrong idea of its effectiveness.

A proper medical trial should ideally obey the following principles:

1. *Controls*: If we want to investigate the effect of statins on a population, we can't just give statins to a few people, and then, if they don't have a heart attack, claim this was due to the pill (regardless of the websites that use this form of anecdotal reasoning to market their products). We need an intervention group, who will be given statins, and a **control group** who will be given sugar pills or **placebos**.

2. *Allocation of treatment*: It is important to compare like with like, so the treatment and comparison groups have to be as similar as possible. The best way to ensure this is by randomly allocating participants to be treated or not, and then seeing what happens to them – this is known as a **randomized controlled trial (RCT)**. Statin trials do this with enough people so that the two groups should be similar in all factors that could otherwise influence the outcome, including – and this is critically important – *those factors that we don't know about*. These studies can be huge: in the UK Heart Protection Study carried out in the late 1990s, 20,536 people at raised risk of a heart attack or stroke were randomly allocated to take either 40 mg of simvastatin daily or a dummy tablet.[3]

3. *People should be counted in the groups to which they were allocated*: The people allocated to the 'statin' group in the Heart Protection Study (HPS) were included in the final analysis *even if they did not take their statins*. This is known as the **'intention to treat'** principle, and can

seem rather odd. It means that the final estimate of the effect of statins really measures the effect of being prescribed statins rather than actually taking them. In practice, of course, people will be strongly encouraged to take the tablets throughout the study, although after five years in the HPS 18% of those allocated a statin had stopped taking them, while as many as 32% of those initially allocated to a placebo tablet actually started taking statins during the trial. Since these people who switch treatments tend to muddy the difference between the groups, we might expect the apparent effect in an 'intention-to-treat' analysis to be less than the effect of actually taking the drug.

4. *If possible, people should not even know which group they are in*: In the statin trials, both the real statins and the placebo pills looked identical and so the participants were **blinded** to the treatment they were taking.

5. *Groups should be treated equally*: If the group allocated to statins were invited back for more frequent hospital appointments, or examined more carefully, it would be impossible to separate the benefits of the drug from the benefits of increased general care. In the HPS, staff doing the follow-up clinics did not know whether the patients were taking a real statin or a placebo, so they were also blinded to the allocated treatment.

6. *If possible, those assessing the final outcomes should not know which group the subjects are in*: If a doctor believes that a treatment works, they may exaggerate benefit for the treatment group through an unconscious bias.

7. *Measure everyone*: Every effort must be made to follow everyone up, as people who drop out of the study might, for example, have done so because of the drug's side effects. The HPS had a remarkable 99.6% complete follow-up at five years, with the results shown in Table 4.1.

Those who were allocated to the statin group clearly had better health outcomes on average, and since patients were randomized and otherwise treated identically, this can be assumed to be a causal effect due to being prescribed statins. But we have seen that many people did not actually adhere to the treatment they had been allocated, and this leads to some dilution of the difference between the groups: the HPS researchers estimate the true effect of actually taking statins is about 50% higher than shown in Table 4.1.

Two final key points:

8. *Don't rely on a single study*: A single statin trial may tell us that the drug worked in a particular group in a particular place, but robust conclusions require multiple studies.

9. *Review the evidence systematically*: When looking at multiple trials, make sure to include every study that has been done, and so create what is known as a systematic review. The results may then be formally combined in a **meta-analysis**.

For example, a recent systematic review put together evidence from twenty-seven randomized trials of statins, which included more than 170,000 people at lower risk of cardiovascular disease.[4] But rather than focusing on the difference between the groups allocated to taking statins and controls, they

Event	Percentage in 10,267 people allocated placebo	Percentage in 10,269 people allocated statin	% (relative) risk reduction in those allocated statins
Heart attack	11.8	8.7	27%
Stroke	5.7	4.3	25%
Death from any cause	14.7	12.9	13%

Table 4.1
The outcomes at five years in the Heart Protection Study, according to treatments allocated to patients. The absolute reduction in the risk of a heart attack was 11.8 − 8.7 = 3.1%. So out of 1,000 people taking a statin, around 31 heart attacks were prevented – this means that around 30 people had to take a statin for five years to prevent one heart attack.

instead estimated the effect of reducing LDL. Essentially they assume the effect of a statin is achieved through changing blood lipids, and based their calculation on the average reduction in LDL seen in each trial, which allows for any non-compliance with allocated treatment. With this extra assumption about the mechanism by which statins benefit our health, they could estimate the effect of actually taking a statin, which they concluded was a 21% reduction in major vascular events per 1 mmol/L (millimoles per litre) reduction in LDL cholesterol. Which is enough for me to keep taking my tablets.*

We have ignored the possibility that any observed relationship is not causal at all, but simply the result of chance. Most drugs on the market have only moderate effects, and only help a minority of people who take them, and their overall benefit can only be reliably detected by large, meticulous, randomized trials. Statin trials are huge, especially when put together in a meta-analysis, which means that the results discussed here cannot be put down to chance variation. (We shall see how to check this in Chapter 10.)

---

* For people with my baseline risk and no previous disease, they estimate a 25% reduction in risk of serious vascular events per 1 mmol/L reduction in LDL. My LDL went down by 2 mmol/L after I started statins, so this should mean my daily pill changes my annual risk of a heart attack or stroke by a factor of around $0.75 \times 0.75 = 0.56$, or equivalently a 44% reduction of my risk. As I had roughly a 13% chance of having a heart attack or stroke in ten years, taking a statin would reduce this to 7%. This means that my being prescribed statins is worthwhile – and even better if I actually take them.

Is prayer effective?

The list of principles for RCTs is not new: they were nearly all introduced in 1948 in what is generally considered the first proper clinical trial. This was of streptomycin, a drug prescribed for tuberculosis. It was bold to randomly allocate patients to either receive or go without this potentially life-saving treatment, but the decision was helped by the fact that there was not enough of the drug for everyone at the time in the UK, and so random allocation seemed a fair and ethical way to decide who should get it. But after all this time, and thousands of RCTs, it can still come as a surprise to the public that medical decisions about what treatment an individual is recommended, even ones as dramatic as whether to have a radical mastectomy or a lumpectomy for breast cancer, have essentially been decided by the flip of a coin (even if it is a metaphorical coin embodied in a computer random number generator).*

In practice the process of allocating treatments in trials is generally more complex than simple randomization case by case, since we want to make sure that all types of people are equally represented in the groups getting different treatments. For example, we may want to have roughly the same number of high-risk older people to get statins and get placebos. This idea came from agricultural experiments, where many of the ideas of randomized trials originated,

---

* It may be even more surprising, and heartening, that so many people had agreed to be part of a trial purely for the benefit of future patients.

largely driven by the work of Ronald Fisher (of whom more later). A large field, for example, would be divided up into individual plots, and then each plot would be randomly allocated to a different fertilizer, just like people being randomly allocated a medical treatment. But parts of the field might be systematically different due to drainage, shade and so on, and so first the field would be divided up into 'blocks' containing plots which were roughly similar. Randomization was then organized in a way that guaranteed that each block contained equal numbers of plots given each fertilizer, which would mean, say, that the treatments were balanced within boggy areas.

For example, I once worked on a randomized trial comparing alternative methods of repairing hernias: standard 'open' surgery versus laparoscopic or 'keyhole' surgery. It was suspected that the skill of the team might increase during the trial, and so it was essential that the two treatments were balanced at all times as the trial progressed. I therefore blocked the sequence of patients into groups of 4 and 6, and made sure patients were randomized equally to each treatment within each block. Back in those days the treatments were printed on little slips of paper, which I folded and placed in opaque numbered brown envelopes. I remember watching patients lying on the pre-op trolley, with no idea which treatment they were going to get, while the anaesthetist opened the envelope to reveal what was going to happen to them, and in particular whether they were going to go home with one large scar or a set of punctures.

Randomized trials became the gold standard for testing new medical treatments, and are now increasingly used to estimate the effects of new policies in education and policing.

For example, the UK Behavioural Insights Team randomly allocated half of students retaking GCSE Mathematics or English to nominate someone to receive regular text messages that encouraged them to support the student in their studies – the students with a 'study supporter' had a 27% higher pass rate. The same team also observed a variety of positive effects in a randomized trial of body-worn video cameras for police officers, such as fewer people being unnecessarily stopped and searched.[5]

There have even been studies to determine the effectiveness of prayer. For example, the Study of the Therapeutic Effects of Intercessory Prayer (STEP) randomly allocated over 1,800 cardiac bypass patients into three groups: patients in Groups 1 and 2 were prayed for and not prayed for, respectively, but did not know which was the case, while Group 3 knew they were being prayed for. The only apparent effect was a small *increase* in complications in the group that knew they were being prayed for: one of the researchers commented, 'It may have made them uncertain, wondering, "Am I so sick they had to call in their prayer team?".'[6]

The main recent innovation in randomized experimentation concerns 'A/B' testing in web design, in which users are (unknowingly) directed to alternative layouts for web pages, and measurements made of time spent on pages, click-throughs to advertisements, and so on. A series of A/B tests can rapidly lead to an optimized design, and the huge sample sizes mean that even small, but still potentially profitable, effects can be reliably detected. This has meant an entirely new community has had to learn about trial design, including the perils of making multiple comparisons that we will come to in Chapter 10.

What Do We Do When We Can't Randomize?

Why do old men have big ears?

It is easy for researchers to randomize if all they have to do is change a website: there is no effort to recruit participants since they don't even know they are the subjects of an experiment, and there is no need to get ethical approval to use them as guinea pigs. But randomization is often difficult and sometimes impossible: we can't test the effect of our habits by randomizing people to smoke or eat unhealthy diets (even though such experiments are performed on animals). When the data does not arise from an experiment, it is said to be observational. So often we are left with trying as best we can to sort out correlation from causation by using good design and statistical principles applied to observational data, combined with a healthy dose of scepticism.

The issue of old men's ears might be rather less important than some of the topics in this book, but illustrates the need for choosing study designs that are appropriate for answering questions. Taking a problem-solving approach based on the PPDAC cycle, the Problem is that, certainly based on my personal observation, old men often seem to have big ears. Why could this be? An obvious Plan is to see whether, in the contemporary population, age is correlated with adult ear-length. It turns out that groups of medical researchers in the UK and Japan have collected Data in such a **cross-sectional** study: their Analysis showed a clear positive correlation, and their Conclusions were that ear-length was associated with age.[7]

The challenge is then to try to explain this association. Do ears carry on growing with age? Or did people who are old now always have bigger ears, and something has happened over the last decades to make more recent generations have smaller ears? Or is it that men with smaller ears simply die earlier for some reason; there is a traditional Chinese belief that big ears predict a longer life. Some imagination is required to think of what kind of studies could test these ideas. A **prospective cohort study** would follow young men through their lives, measuring their ears to check if they grew, or if those with smaller ears died earlier. This would take rather a long time, and so an alternative **retrospective cohort study** could take men who are old now, and try and work out whether their ears had grown, perhaps using past photographic evidence. A **case-control study** could take men who died, find men who are still alive who matched them in their age and other factors that are known to predict longevity, and see if the survivors had bigger ears.*

And so the problem-solving cycle would start again.

What Can We Do When We Observe an Association?

This is where some statistical imagination is called for, and it can be an enjoyable exercise to guess the reasons why an observed correlation might be spurious. Some are fairly easy: the close correlation between mozzarella consumption and

---

* Sadly, it is unlikely that any of these proposals are likely to attract funding.

civil engineers is presumably because both measures have been increasing over time. Similarly any correlation between ice-cream sales and drownings is due to both being influenced by the weather. When an apparent association between two outcomes might be explained by some observed common factor that influences both, this common cause is known as a **confounder**: both the year and weather are potential confounders since they can be recorded and considered in an analysis.

The simplest technique for dealing with confounders is to look at the apparent relationship within each level of the confounder. This is known as **adjustment**, or stratification. So for example we could explore the relationship between drownings and ice-cream sales on days with roughly the same temperature.

But adjustment can produce some paradoxical results, as shown by an analysis of acceptance rates by gender at Cambridge University. In 1996 the overall acceptance rate to study five academic subjects in Cambridge was slightly higher for men (24% of 2,470 applicants) than it was for women (23% of 1,184 applicants). The subjects were all in what we today call STEM (science, technology, engineering and medicine) subjects, which have historically been studied predominantly by men. Was this a case of gender discrimination?

Take a careful look at Table 4.2. Although overall the acceptance rate was higher for men, the acceptance rate in each subject individually was higher for women. How can this apparent paradox occur? The explanation is that the women were more likely to apply for the more popular and therefore more competitive subjects with the lowest acceptance rate, such as medicine and veterinary medicine, and tended not to apply to engineering, which has a higher acceptance rate. In

	Women			Men		
	Applied	Accepted	%	Applied	Accepted	%
Computer Science	26	7	27%	228	58	25%
Economics	240	63	26%	512	112	22%
Engineering	164	52	32%	972	252	26%
Medicine	416	99	24%	578	140	24%
Veterinary Medicine	338	53	16%	180	22	12%
TOTAL	1,184	274	23%	2,470	584	24%

Table 4.2
Illustration of Simpson's Paradox using admission data for
Cambridge in 1996. Overall, the acceptance rate was higher
for men. But in each subject the acceptance rate was higher
for women.

this case, therefore, we might conclude that there is no evidence of discrimination.

This is known as **Simpson's paradox**, which occurs when the apparent direction of an association is reversed by adjusting for a confounding factor, requiring a complete change in the apparent lesson from the data. Statisticians revel in finding real-life examples of this, each further reinforcing the caution required in interpreting observational data. Nevertheless, it shows the insights gained by splitting data according to factors that may help explain observed associations.

> Does having a nearby Waitrose put £36,000 on the value of your house?

The claim that a nearby Waitrose 'adds £36,000 to house price' was credulously reported by the British media in 2017.[8] But this was not a study of the change in house prices after a store opened, and Waitrose certainly did not experimentally randomize the placement of their new stores: it was simply a correlation between house prices and the closeness of supermarkets, particularly upscale ones like Waitrose.

The correlation almost certainly reflects Waitrose's policy of opening stores in wealthier locations, and is therefore a fine example of the actual chain of causation being the precise opposite of what has been claimed. This is known, unsurprisingly, as **reverse causation**. More serious examples occur in studies examining the relationship between drinking alcohol and health outcomes, which generally find that non-drinkers have substantially higher death rates than moderate drinkers. How can this possibly make sense, given what we know about the impact of alcohol on the liver for example? This relationship has been partially attributed to reverse causation – those people who are more likely to die do not drink because they are ill already (possibly through excessive drinking in the past). More careful analyses now exclude ex-drinkers, and also ignore adverse health events in the first few years of the study, since these may be due to pre-existing conditions. Even with these exclusions, some over-all health benefit from moderate drinking appears to remain, although it is deeply contested.

Another amusing exercise is to try to invent a narrative of reverse causation for any statistical claim based on correlation alone. My favourite is a study finding a correlation between US teenagers' consumption of carbonated soft drinks and their tendency towards violence: while a newspaper reported this as 'Fizzy Drinks Make Teenagers Violent',[9] perhaps it is just as plausible that being violent works up a thirst? Or more plausibly we could think of some common factors that might influence both, such as membership of a particular peer-group. Potential common causes that we do not measure are known as **lurking factors**, since they remain in the background, are not included in any adjustment, and are waiting to trip up naïve conclusions from observational data.

Here are some more examples of how easy it might be to believe a causal link, when some other factor is influencing events:

- Many children are diagnosed with autism soon after being vaccinated. Does vaccination cause autism? No, these are events that happen at around the same age and inevitably there are some coincidental close occurrences.

- Out of the total number of people who die each year, a smaller proportion are left-handed than in the general population. Does that mean that left-handers live longer? No, this happens because people who are dying now were born in an era when children were forced to change to being right-handed, and so there are simply fewer older left-handers.[10]

- The average age at which popes die is older than that of the general population. Does this mean that being a pope helps you live longer? No, popes are selected from a group who have not died young (otherwise they could not be candidates).[11]

The myriad ways we can be caught out might encourage the idea that we can never conclude causation from anything other than a randomized experiment. But, perhaps ironically, this view was counteracted by the man responsible for the first modern randomized clinical trial.

Can We Ever Conclude Causation from Observational Data?

Austin Bradford Hill was a brilliant British applied statistician who was at the forefront of two world-changing scientific advances: he designed the streptomycin clinical trial mentioned earlier in the chapter, which essentially set the standards for all subsequent RCTs, and with Richard Doll in the 1950s he led the research that eventually confirmed the link between smoking and lung cancer. In 1965 he set out a list of criteria that needed to be considered before concluding that an observed link between an **exposure** and an outcome was causal, where an exposure might comprise anything from chemicals in the environment to habits such as smoking or lack of exercise.

These criteria have been subsequently much debated, and the version shown below was developed by Jeremy Howick and colleagues, separated into what they call direct, mechanistic and parallel evidence.[12]

Direct evidence:

1. The size of the effect is so large that it *cannot be explained by plausible confounding*.
2. There is *appropriate temporal and/or spatial proximity*, in that cause precedes effect and effect occurs after a plausible interval, and/or cause occurs at the same site as the effect.
3. *Dose responsiveness and reversibility*: the effect increases as the exposure increases, and the evidence is even stronger if the effect reduces upon reduction of the dose.

Mechanistic evidence:

4. There is a *plausible mechanism of action*, which could be biological, chemical, or mechanical, with external evidence for a 'causal chain'.

Parallel evidence:

5. The effect fits with what is known already.
6. The effect is found when the study is replicated.
7. The effect is found in similar, but not identical, studies.

These guidelines might enable causation to be determined from anecdotal evidence, even in the absence of a randomized trial. For example, mouth ulcers have been observed to occur after aspirin is rubbed within the mouth, say to relieve tooth pain. The effect is dramatic (obeys guideline 1), occurs where rubbed (2), is a plausible response to an acidic compound (4), is not contradicted by current science and is similar to the known effect of aspirin in causing stomach ulcers (5), and has been repeatedly observed in multiple patients (6). So five out of seven guidelines are satisfied, the remaining two have not been tested, and so it is reasonable to conclude this is a genuine adverse reaction to the drug.

The Bradford Hill criteria apply to general scientific conclusions for populations. But we may also be interested in individual cases, say in civil litigation where courts need to decide whether a particular exposure (say the asbestos encountered in a job) caused a negative outcome in a specific person (say John Smith's lung cancer). It can never be established with absolute certainty that the asbestos was the cause of the cancer, since it cannot be proved that the cancer would not have occurred without the exposure. But some courts have accepted that, on the 'balance of probabilities', a direct causal link has been established if the relative risk associated with the exposure is greater than two. But why two?

Presumably the reasoning behind this conclusion is as follows:

1. Suppose that, in the normal run of things, out of 1,000 men like John Smith, 10 would get lung cancer. If

asbestos more than doubles the risk, then if these 1,000 men had been exposed to asbestos, then perhaps 25 would have developed lung cancer.

2. So of those exposed to asbestos who go on to develop lung cancer, less than half would have got lung cancer if they had not been exposed.

3. So more than half of the lung cancers in this group will have been caused by the asbestos.

4. Since John Smith is one of this group of people, then on the balance of probabilities his lung cancer was caused by the asbestos.

This kind of argument has led to a new area of study known as **forensic epidemiology**, which tries to use evidence derived from populations to draw conclusions about what might have caused individual events to occur. In effect this discipline has been forced into existence by people seeking compensation, but this is a very challenging area for statistical reasoning about causation.

The appropriate handling of causation still remains contested within the field of statistics, whether it concerns pharmaceuticals or big ears, and without randomization it is rare to be able to draw confident conclusions. One imaginative approach takes advantage of the fact that many genes are spread essentially at random through the population, so it is as if we have been randomized to our specific version at conception. This is known as Mendelian randomization, after Gregor Mendel, who developed the modern idea of genetics.[13]

Other advanced statistical methods have been developed to try to adjust for potential confounders and so to get closer to an estimate of the actual effect of the exposure, and these are largely based on the important idea of regression analysis. And for this we must acknowledge, yet again, the fertile imagination of Francis Galton.

Summary

- Causation, in the statistical sense, means that when we intervene, the chances of different outcomes are systematically changed.
- Causation is difficult to establish statistically, but well-designed randomized trials are the best available framework.
- Principles of blinding, intention-to-treat and so on have enabled large-scale clinical trials to identify moderate but important effects.
- Observational data may have background factors influencing the apparent observed relationships between an exposure and an outcome, which may be either observed confounders or lurking factors.
- Statistical methods exist for adjusting for other factors, but judgement is always required as to the confidence with which causation can be claimed.

Modelling Relationships Using Regression

The ideas in previous chapters allow us to visualize and summarize a single set of numbers, and also to look at associations between pairs of variables. These basic techniques can take us a remarkably long way, but modern data will generally be a lot more complex. There will often be a list of possibly related variables, one of which we are particularly interested in explaining or predicting, whether it is an individual's risk of cancer or a country's future population. In this chapter we meet the important idea of a **statistical model**, which is a formal representation of the relationships between variables, which we can use for the desired explanation or prediction. This inevitably means introducing some mathematical ideas, but the basic concepts should be clear without using algebra.

But first we return to Francis Galton. He had the classic Victorian gentleman scientist's obsessive interest in collecting data, and eliciting the wisdom of crowds about the weight of an ox is only one example. He used his observations to make weather forecasts, to assess the efficacy of prayer and even to compare the relative beauty of young women in different parts of the country.* He also shared his cousin

---

* According to Galton, 'I found London to rank the highest for beauty: Aberdeen lowest.'

Charles Darwin's fixation on inheritance, and set out to investigate the way that personal characteristics change between generations. He was particularly interested in the following question:

Using their parents' heights, how can we predict an adult offspring's height?

In 1886 Galton reported the heights of a large group of parents and their adult children, and summary statistics for the majority of the data are shown in Table 5.1.[1] Galton's sample had similar heights to contemporary adults (the average heights for adult women and men in the UK in 2010 was reported to be 63 and 69 inches respectively), which suggests his subjects were well-nourished and of higher socioeconomic status.

Figure 5.1 shows a scatter-plot of 465 sons' heights against their fathers' heights. The heights of fathers and sons are clearly correlated, with a Pearson correlation of 0.39. What if we wanted to predict a son's height from his father's? We might start by choosing a straight line to make our predictions, since that will enable us, for any father's height, to calculate a prediction for their son's stature. Our immediate intuition might be to use a diagonal line of 'equality', so that an adult son is predicted to have the same height as his father. But it turns out we can improve on this choice.

For any straight line we choose, each data-point will give rise to a **residual** (the vertical dashed lines on the plot), which is the size of the error were we to use the line to predict a son's height from his father's. We want a line that makes

	Number	Mean	Median	Standard deviation
Mothers	197	64.0	64.0	2.4
Fathers	197	69.3	69.5	2.6
Daughters	433	64.1	64.0	2.4
Sons	465	69.2	69.2	2.6

Table 5.1
Summary statistics of recorded heights (in inches) of 197 sets
of parents and their adult children recorded by Galton in 1886.
For reference, 64 inches is 1.63 metres, 69 inches is 1.75 metres.
Even without plotting the data, the closeness of the mean and
median suggests a symmetric data distribution.

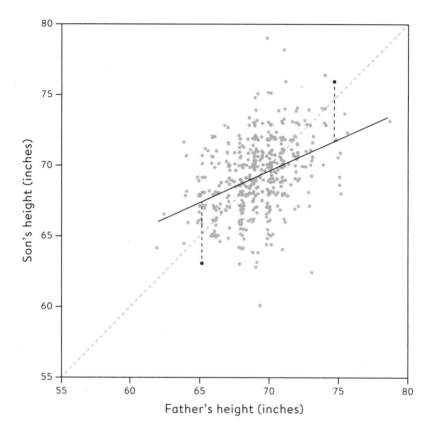

Figure 5.1

Scatter of heights of 465 fathers and sons from Galton's data (many fathers are repeated since they have multiple sons). A jitter has been added to separate the points, and the diagonal dashed line represents exact equality between son and father's heights. The solid line is the standard 'best-fit' line. Each point gives rise to a 'residual' (dashed line), which is the size of the error were we to use the line to predict a son's height from his father's.

these residuals small, and the standard technique is to choose a **least-squares** fitted line, for which the sum of the squares of the residuals is smallest.* The formula for this line is straightforward (see the Glossary), and was developed by both French mathematician Adrien-Marie Legendre and Carl Friedrich Gauss at the end of the eighteenth century. The line is generally known as the 'best-fit' prediction we can make about a son's height, from knowing his father's.

The least-squares prediction line in Figure 5.1 goes through the middle of the cloud of points, representing the mean values for the heights of fathers and sons, but does not follow the diagonal line of 'equality'. It is clearly lower than the line of equality for fathers who are taller than average, and higher than the line of equality for fathers who are shorter than average. This means that tall fathers tend to have sons who are slightly shorter than them, while shorter fathers have slightly taller sons. Galton called this 'regression to mediocrity', whereas now it is known as **regression to the mean.** The phenomenon also holds for mothers and daughters: taller mothers tend to have daughters who are shorter than them, and shorter mothers tend to have taller daughters. This explains the origin of the term in this chapter's title: eventually any process of fitting lines or curves to data came to be called 'regression'.

In basic regression analysis the dependent variable is the quantity that we want to predict or explain, usually forming

---

* It would be possible to fit a line that minimized the sum of the absolute values of the residuals, rather than the sum of their squares, but this would be almost impossible without a modern computer.

the vertical y-axis of a graph – this is sometimes known as the response variable. While the independent variable is the quantity that we use for doing the prediction or explanation, usually forming the horizontal x-axis of a graph, and sometimes known as the explanatory variable. The gradient is also known as the **regression coefficient**.

Table 5.2 shows the correlations between parent and offspring heights, and the gradients of regression lines.[*] There is a simple relationship between the gradients, the Pearson correlation coefficient and the standard deviations of the variables.[†] In fact if the standard deviations of the independent and dependent variables are the same, then the gradient is simply the Pearson correlation coefficient, which explains their similarity in Table 5.2.

The meaning of these gradients depends completely on our assumptions about the relationship between the variables being studied. For correlational data, the gradient indicates how much we would expect the dependent variable to change, on average, if we observe a one unit difference for the independent variable. For example, if Alice is one inch taller than Betty, we would predict Alice's adult daughter to be 0.33 inches taller than Betty's adult daughter. Of course we would not expect this prediction to match their true difference in heights precisely, but it is the best guess we can make with the data available.

If, however, we assumed a *causal* relationship then the

---

[*] For example, we would predict a daughter's height by the formula: average height of all daughters + 0.33 × (mother's height − average height of all mothers).

[†] See the Glossary entry for least-squares.

	Pearson Correlation	Gradient of regression of offspring on parent
Mothers and daughters	0.31	0.33
Fathers and sons	0.39	0.45

Table 5.2
Correlations between heights of adult children and parent of the same gender, and gradients of the regression of the offspring's on the parent's height.

gradient has a very different interpretation – it is the change we would expect in the dependent variable were we to intervene and change the independent variable to a value one unit higher. This is definitely not the case for heights since they cannot be altered by experimental means, at least for adults. Even with the Bradford Hill criteria outlined above, statisticians are generally reluctant to attribute causation unless there has been an experiment, although computer scientist Judea Pearl and others have made great progress in setting out the principles for building causal regression models from observational data.[2]

Regression Lines Are Models

The regression line we fitted between fathers' and sons' heights is a very basic example of a statistical model. The US Federal Reserve define a model as a 'representation of some aspect of the world which is based on simplifying assumptions': essentially some phenomenon will be represented mathematically, generally embedded in computer software, in order to produce a simplified 'pretend' version of reality.[3]

Statistical models have two main components. First, a mathematical formula that expresses a deterministic, predictable component, for example the fitted straight line that enables us to make a prediction of a son's height from his father's. But the deterministic part of a model is not going to be a perfect representation of the observed world. As we saw in Figure 5.1, there is a big scatter of heights around the regression line, and the difference between what the model predicts, and what actually happens, is the second component of a model and is known as the **residual error** – although it is important to remember that in statistical modelling, 'error' does not refer to a mistake, but the inevitable inability of a model to exactly represent what we observe. So in summary, we assume that

observation = deterministic model + residual error.

This formula can be interpreted as saying that, in the statistical world, what we see and measure around us can be considered as the sum of a systematic mathematical idealized form plus some random contribution that cannot yet be explained. This is the classic idea of the **signal and the noise**.

Do speed cameras reduce accidents?

This section contains a simple lesson: just because we act, and something changes, it doesn't mean we were responsible for the result. Humans seem to find this simple truth difficult to grasp – we are always keen to construct an explanatory narrative, and even keener if we are at its centre. Of course sometimes this interpretation is true – if you flick a switch, and the light comes on, then you are usually responsible. But sometimes your actions are clearly not responsible for an outcome: if you don't take an umbrella, and it rains, it is not your fault (although it may feel that way). But the consequences of many of our actions are less clear-cut. Suppose you have a headache, take an aspirin, and your headache goes away. How do you know it wouldn't have gone away even if you had not taken a tablet?

We have a strong psychological tendency to attribute change to intervention, and this makes before-and-after comparisons treacherous. A classic example concerns speed cameras, which tend to get put in places that have recently experienced accidents. When the accident rate subsequently goes down, this change is then attributed to the presence of the cameras. But would the accident rates have gone down anyway?

Strings of good (or bad) luck do not go on for ever, and eventually things settle back down – this can also be considered as regression-to-the-mean, just like tall fathers tending to have shorter sons. But if we believe these runs of good or bad fortune represent a constant state of affairs, then we will wrongly attribute the reversion to normal as the consequence of any intervention we have made. Perhaps this all seems rather obvious, but this simple idea has remarkable ramifications, such as:

- Football managers who get sacked after a string of losses, only to find their successors getting credit for the return to normal.
- Active fund managers dropping in performance, after being tipped (and perhaps getting large bonuses) after a couple of good years.
- The 'Curse of *Sports Illustrated*', in which athletes get featured on the cover of a prominent magazine following a series of achievements, only to subsequently have their performance plummet.

Luck plays a considerable part in the position that sports teams have in their league table, and a consequence of regression-to-the-mean means we would expect teams that do well one year to decline the following year, and those that do badly to improve their position, particularly if the teams are fairly evenly matched. Conversely, if we see this pattern of changes, we might suspect that regression-to-the-mean is operating and not take too much notice of claims about the influence of, say, new training methods.

It is not only sports teams that are ranked in league tables. Take the example of the PISA Global Education Tables, which compare different countries' school systems in mathematics. A change in league table position between 2003 and 2012 was strongly negatively correlated with initial position, meaning that countries at the top tended to go down, and those at the bottom tended to go up. The correlation was −0.60, and some theory shows that if the rankings were complete chance and all that was operating were regression-to-the-mean, the correlation would be expected to be −0.71, not very different from what was observed.[4] This suggests the differences between countries were far less than claimed, and that changes in league position had little to do with changes in teaching philosophy.

Regression-to-the-mean also operates in clinical trials. In the last chapter we saw that randomized trials were needed to evaluate new pharmaceuticals properly, since even people in the control arm showed benefit – the so-called placebo effect. This is often interpreted to mean that just taking a sugar pill (preferably a red one) actually has a beneficial effect on people's health. But much of the improvement seen in people who do not receive any active treatment may be regression-to-the-mean, since patients are enrolled in trials when they are showing symptoms, and many of these would have resolved anyway.

So if we want to know the genuine effect of installing speed cameras in accident black spots, then we should follow the approach used for evaluating pharmaceuticals and take the bold step of randomly allocating speed cameras. When such studies have been conducted, it is estimated that about two-thirds of the apparent benefit from cameras is due to regression-to-the-mean.[5]

Dealing With More Than One Explanatory Variable

Since Galton's early work there have been many extensions to the basic idea of regression, vastly helped by modern computing. These developments include:

- having many explanatory variables
- explanatory variables that are categories rather than numbers
- having relationships that are not straight lines and adapt flexibly to the pattern of the data
- response variables that not continuous variables, such as proportions and counts

As an example of having more than one explanatory variable, we can look at how the height of a son or daughter is related to the height of their father *and* their mother. The scatter of data-points is now in three dimensions and becomes much more difficult to draw on a page, but we can still use the idea of least-squares to work out the formula that best predicts offspring height. This is known as a **multiple linear regression.**[*]

---

[*] The 'linear' refers to the fact that this equation consists of a weighted sum of the explanatory variables, weighted by their regression coefficients, and this is known as a linear model.

When we just had one explanatory variable the relationship with the response variable was summarized by a gradient, which can also be interpreted as a coefficient in a regression equation; this idea can be generalized to more than one explanatory variable.

The results for Galton's families are shown in Table 5.3. How can we interpret the coefficients shown here? First, they are part of a formula that could be used to predict adult offspring height for a particular mother and father.* But they also illustrate the idea of adjustment of an apparent relationship, by taking account of a third, confounding factor.

For example, we saw in Table 5.2 that the gradient when regressing the height of daughters on their mother's height was 0.33 – remember that the gradient of a line fitted to a scatter-plot is the just another name for the regression coefficient. Table 5.3 shows that, if we also allow for the effect of father's height, this coefficient is reduced to 0.30. When predicting a son's height, the regression coefficient for the father is similarly reduced from 0.45 in Table 5.2 to 0.41 in Table 5.3, when the mother's height is taken into account. So the height of a parent has a slightly reduced association with their adult offspring's height, when allowing for the effect of the other parent. This could be due to the fact that taller women tend to marry taller men, so that each parent's height is not a completely independent factor. Overall, the data suggests a one inch difference in a father's height is associated

---

* The explanatory variables have been standardized by subtracting their mean value in the sample. So to predict the height of a son, we would use the formula: 69.2 + 0.33 (mother's height – average height of mothers) + 0.41 (father's height – average height of fathers)

Dependent variable	Intercept (average height of offspring)	Coefficient of multiple regression on mother's height	Coefficient of multiple regression on father's height
Daughter's height	64.1	0.30	0.40
Son's height	69.2	0.33	0.41

Table 5.3
Results of a multiple linear regression relating adult offspring height to that of their mother and father. The 'intercept' is the average height of offspring (Table 5.1). The coefficients of multiple regression indicate the predicted change in adult offspring height for each one-inch change from the average parental height.

with a bigger difference in an adult child's height than a one inch difference in a mother's height. Multiple regression is often used when researchers are interested in one particular explanatory variable, and other variables need to be 'adjusted for' to allow for imbalances.

Let's return to the Swedish study on brain tumours that we saw in Chapter 4 as an example of an inappropriate media interpretation of causation. A regression analysis had the rate of tumours as the dependent, or response, variable, and education as the independent, or explanatory, variable of interest. Other factors entered into the regression included age at diagnosis, calendar year, region of Sweden, marital status and income, all of which were considered to be potential confounding variables. This adjustment for confounders is an attempt to tease out a purer relationship between education and brain tumours, but it can never be wholly adequate. There will always remain the suspicion that some other lurking process might be at work, such as those with higher education seeking better health care and increased diagnoses.

In a randomized trial, there should be no need to adjust for confounders, as the random allocation should guarantee that all factors other than the main treatment should be balanced between groups. But researchers often still carry out a regression analysis anyway, just in case some imbalances have slipped in.

Different Types of Response Variables

Not all data are continuous measurements such as height. In much of statistical analysis, the dependent variables may be the proportions of events that either happen or not (for example, the proportion of people who survive surgery), counts of the numbers of events (for example, how many cancers occur per year in a certain area), or the length of time before an event occurs (for example, years of survival following surgery). Each type of dependent variable has its own form of multiple regression, with a correspondingly different interpretation of the estimated coefficients.[6]

Consider the child heart surgery data discussed in Chapter 2, where Figure 2.5(a) showed the proportions surviving surgery and the number of cases treated in each hospital between 1991 and 1995. The scatter-plot is shown again in Figure 5.2 with a regression line that has been fitted without using the outlying data-point corresponding to Bristol.

While we could have fitted a linear regression line through these points, naïve extrapolation would suggest that if a hospital treated a huge number of cases, their survival would be predicted to be greater than 100%, which is absurd. So a form of regression has been developed for proportions, called **logistic regression**, which ensures a curve which cannot go above 100% or below 0%.

Even without taking Bristol into account, hospitals with more patients had better survival rates, and the logistic regression coefficient (0.001) means the mortality rate is expected to be around 10% lower (relatively) for each additional 100 operations that a hospital conducts on under-1s over a four-year period.* Of course, to use what is now rather a cliché,

---

* The logistic regression coefficient means that the logarithm of the odds of mortality is estimated to decrease by 0.001 per extra patient treated per year, and so it decreases by 0.1 for each extra 100 patients. This corresponds to around a 10% lower risk.

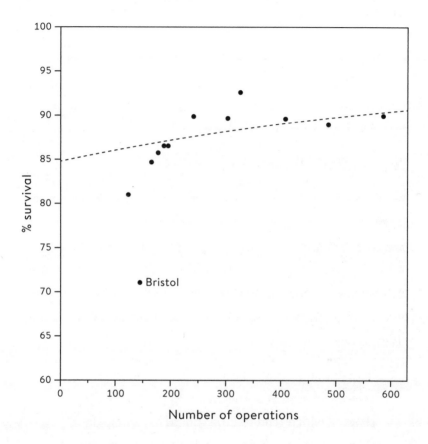

Figure 5.2
Fitted logistic regression model for child heart surgery data for under-1s in UK hospitals between 1991 and 1995. Hospitals treating more patients have better survival. The line is part of a curve that will never reach 100%, and is fitted ignoring the outlying data-point representing Bristol.

correlation does not mean causation, and we cannot conclude that bigger throughput is the reason for the better performance: as we mentioned previously, there could even be reverse causation, with hospitals with a good reputation attracting more patients.

This was a controversial finding when it was released in 2001, and has contributed to prolonged, and still unresolved, disputes about how many hospitals in the UK should conduct this form of surgery.

Beyond Basic Regression Modelling

The techniques outlined in this chapter have worked remarkably well since their introduction more than a century ago. But both the availability of large amounts of data and the extraordinary increase in computing power have allowed far more sophisticated models to be developed. Very broadly, four main modelling strategies have been adopted by different communities of researchers:

- Rather simple mathematical representations for associations, such as the linear regression analyses in this chapter, which tend to be favoured by statisticians.
- Complex deterministic models based on scientific understanding of a physical process, such as those used in weather forecasting, which are intended to realistically represent underlying mechanisms, and which are generally developed by applied mathematicians.
- Complex algorithms used to make a decision or prediction that have been derived from an analysis of huge numbers of past examples, for example to

recommend books you might like to buy from an online retailer, and which come from the world of computer science and **machine learning**. These will often be 'black boxes' in the sense that they may make good predictions, but their internal structure is somewhat inscrutable – see the next chapter.

- Regression models that claim to reach causal conclusions, as favoured by economists.

These are huge generalizations, and fortunately professional barriers are breaking down and we shall see later that a more ecumenical approach to modelling is developing. But whatever the strategy adopted, common issues arise when building and using a model.

A good analogy is that a model is like a map, rather than the territory itself. And we all know that some maps are better than others: a simple one might be good enough to drive between cities, but we need something more detailed when walking through the countryside. The British statistician George Box has become famous for his brief but invaluable aphorism: 'All models are wrong, some are useful.' This pithy statement was based on a lifetime spent bringing statistical expertise to industrial processes, which led Box to appreciate both the power of models, but also the danger of actually starting to believe in them too much.

But these cautions are easily forgotten. Once a model becomes accepted, and especially when it is out of the hands of those who created it and understand its limitations, then it can start acting as a sort of oracle. The financial crisis of 2007–2008 has to a large extent been blamed on the exaggerated trust placed in complex financial models used to determine the risk of, say, bundles of mortgages. These models assumed only a moderate correlation between mortgage failures, and worked well while the property market was booming. But when conditions changed and mortgages starting failing, they tended to fail in droves: the models grossly underestimated the risks due to the correlations turning out to be far higher than supposed. Senior managers simply did not realize the frail basis on which these models were built, losing track of the fact that models are simplifications of the real world – they are *the maps not the territory*. The result was one of the worst global economic crises in history.

Summary

- Regression models provide a mathematical representation between a set of explanatory variables and a response variable.
- The coefficients in a regression model indicate how much we expect the response to change when the explanatory variable is observed to change.
- Regression-to-the-mean occurs when more extreme responses revert to nearer the long-term average, since a contribution to their previous extremeness was pure chance.
- Regression models can incorporate different types of response variable, explanatory variables and non-linear relationships.
- Caution is required in interpreting models, which should not be taken too literally: 'All models are wrong, but some are useful.'

CHAPTER 6

Algorithms, Analytics and Prediction

The emphasis so far in this book has been on how statistical science can help us to understand the world, whether it is working out the potential harm of eating bacon sandwiches, or the relationship between the height of parents and offspring. This is essentially scientific research, to work out what is really going on and what, in the terms introduced in the last chapter, is just residual error to be treated as unavoidable variability that cannot be modelled.

But the basic ideas of statistical science still hold when we are trying to solve a practical rather than a scientific problem. The basic desire to find the signal in the noise is just as relevant when we just want a method that will help in a particular decision faced in our daily lives. The theme behind this chapter is that such practical problems can be tackled by using past data to produce an algorithm, a mechanistic formula that will automatically produce an answer for each new case that comes along with either no, or minimal, additional human intervention: essentially, this is 'technology' rather than science.

There are two broad tasks for such an algorithm:

- Classification (also known as discrimination or **supervised learning**): to say what kind of situation we're facing. For example, the likes and dislikes of an online customer, or whether that object in a robot's vision is a child or a dog.

- Prediction: to tell us what is going to happen. For example, what the weather will be next week, what a stock price might do tomorrow, what products that customer might buy, or whether that child is going to run out in front of our self-driving car.

Although these tasks differ in whether they are concerned with the present or the future, they both have the same underlying nature: to take a set of observations relevant to a current situation, and map them to a relevant conclusion. This process has been termed **predictive analytics**, but we are verging into the territory of **artificial intelligence (AI)**, in which algorithms embodied in machines are used either to carry out tasks that would normally require human involvement, or to provide expert-level advice to humans.

'Narrow' AI refers to systems that can carry out closely prescribed tasks, and there have been some extraordinarily successful examples based on machine learning, which involves developing algorithms through statistical analysis of large sets of historical examples. Notable successes include speech recognition systems built into phones, tablets and computers; programs such as Google Translate which know little grammar but have learned to translate text from an immense published archive; and computer vision software that uses past images to 'learn' to identify, say, faces in photographs or other cars in the view of self-driving vehicles. There has also been spectacular progress in systems playing games, such as the DeepMind software learning the rules of computer games and becoming an expert player, beating world-champions at chess and Go, while IBM's Watson has beaten competing humans in general knowledge quizzes. These systems did not begin by trying to encode human expertise and knowledge. They started with a vast number of examples, and learned through trial and error rather like a naïve child, even by playing themselves at games.

But again we should emphasize that these are technological systems that use past data to answer immediate practical questions, rather than scientific systems that seek to understand how the world works: they are to be judged solely on how well they carry out the limited task at hand, and, although the form of the learned algorithms may provide some insights, they are not expected to have imagination or have super-human skills in everyday life. This would require 'general' AI, which is both beyond the content of this book and, at least at present, beyond the capacity of machines.

Ever since formulae for calculating insurance and annuities were developed by Edmund Halley in the 1690s, statistical science has been concerned with producing algorithms to help in human decisions. The modern development of data science continues that tradition, but what has changed in recent years is the scale of the data being collected and the imaginative products being developed: so-called 'big data'.

Data can be 'big' in two different ways. First, in the number of examples in the database, which may be individual people but could be stars in the sky, schools, car rides or social media posts. The number of examples is often given the label n, and in my early days n was 'big' if it was anything more than 100, but now there may be data on many millions or billions of cases.

The other way that data can be 'big' is by measuring many characteristics, or features, on each example. This quantity is often known as p, perhaps denoting parameters. Thinking again back to my statistical youth, p used to be generally less than 10 – perhaps we knew a few items of an individual's medical history. But then we started having access to millions of that person's genes, and genomics became a small n, large p problem, where there was a huge amount of information about a relatively small number of cases.

And now we have entered the era of large n, large p problems, in which there are vast numbers of cases, each of which may be very complex – think of the algorithms that are analysing all the posts, likes and dislikes of each of the billions of Facebook subscribers to decide what sort of adverts and news to feed them.

These are exciting new challenges which have brought waves of new people into data science. But, to refer yet again to the warning at the start of this book, these trough-loads of data do not speak for themselves. They need to be handled with care and skill if we are to avoid the many potential pitfalls of using algorithms naïvely. We shall see some classic disasters in this chapter, but first we need to consider the fundamental problem of boiling the data down into something useful.

Finding Patterns

One strategy for dealing with an excessive number of cases is to identify groups that are similar, a process known as clustering or **unsupervised learning**, since we have to learn about these groups and are not told in advance that they exist. Finding these fairly homogeneous clusters can be an end in itself, for example by identifying groups of people with similar likes and dislikes, which then can be characterized, given a label, and algorithms built for classifying future cases. The clusters that have been identified can then be fed appropriate film recommendations, advertisements, or political propaganda, depending on the motivation of the people building the algorithm.

Before getting on with constructing an algorithm for classification or prediction, we may also have to reduce the raw data on each case to a manageable dimension due to excessively large p, that is too many features being measured on each case. This process is known as **feature engineering**. Just think of the number of measures that could be made on a human face, which may need to be reduced to a limited number of important features that can be used by facial recognition software to match a photograph to a database. Measures that lack value for prediction or classification may be identified by data visualization or regression methods and then discarded, or the numbers of features may be reduced by forming composite measures that encapsulate most of the information.

Recent developments in extremely complex models, such as those labelled as **deep learning**, suggest that this initial stage of data reduction may not be necessary and the total raw data can be processed in a single algorithm.

Classification and Prediction

A bewildering range of alternative methods are now readily available for building classification and prediction algorithms. Researchers used to promote methods which came from their own professional backgrounds: for example statisticians preferred regression models, while computer scientists preferred rule-based logic or 'neural networks' which were alternative ways to try and mimic human cognition. Implementation of any of these methods required specialized skills and software, but now convenient programs allow a menu-driven choice of technique, and so encourage a less partisan approach where performance is more important than modelling philosophy.

As soon as the practical performance of algorithms started to be measured and compared, people inevitably got competitive, and now there are data science contests hosted by platforms such as Kaggle.com. A commercial or academic organization provides a data set for competitors to download: challenges have included detecting whales from sound recordings, accounting for dark matter in astronomical data, and predicting hospital admissions. In each case competitors are provided with a training set of data on which to build their algorithm, and a test set that will decide their performance. A particularly popular competition, with thousands of competing teams, is to produce an algorithm for the following challenge.

Can we predict which passengers survived the sinking of the *Titanic*?

On its maiden voyage, the *Titanic* hit an iceberg and slowly sank on the night of 14/15 April 1912. Only around 700 of more than 2,200 passengers and crew on board got on to lifeboats and survived, and subsequent studies and fictional accounts have focused on the fact that your chances of getting on to a lifeboat and surviving crucially depended on what class of ticket you had.

An algorithm that predicts survival may at first seem an odd choice of Problem within the standard PPDAC cycle, since the situation is hardly likely to arise again, and so is not going to have any future value. But a specific individual provided me with some motivation. In 1912 Francis William Somerton left Ilfracombe in north Devon, close to where I was born and brought up, to go to the US to make his fortune. He left his wife and young daughter behind, and bought a third-class ticket costing £8 1s. for the brand-new *Titanic*. He never made it to New York – his memorial is in Ilfracombe churchyard (Figure 6.1). An accurate predictive algorithm will be able to tell us whether Francis Somerton was unlucky not to survive, or whether his chances were in fact slim.

The Plan is to amass available data and try a range of different techniques for producing algorithms that predict who survived – this could be considered more of a classification than a prediction problem, since the events have already happened. The Data comprise publicly available information on 1,309 passengers on the *Titanic*: potential predictor variables include their full name, title, gender, age, class of travel (first, second, third), how much they paid for their ticket, whether they were part of a family, where they boarded the boat (Southampton, Cherbourg, Queenstown), and limited data

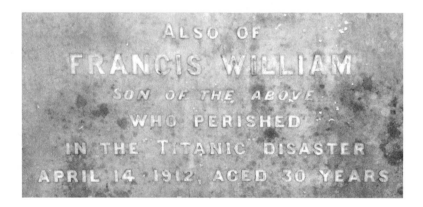

Figure 6.1
The memorial to a Francis William Somerton in the churchyard in
Ilfracombe. It reads, 'Also of Francis William, son of the above, who
perished in the Titanic disaster April 14 1912, aged 30 years'.

on some cabin numbers.[1] The response variable is an indicator for whether they survived (1) or not (0).

For the Analysis, it is crucial to split the data into a training set used to build the algorithm, and a test set that is kept apart and only used to assess performance – it would be serious cheating to look at the test set before we are ready with our algorithm. Like the Kaggle competition, we will take a random sample of 897 cases as our training set, and the remaining 412 individuals will comprise the test set.

This is a real, and hence fairly messy, data set, and some pre-processing is required. Eighteen passengers have missing fare information, and they have been assumed to have paid the median fare for their class of travelling. The number of siblings and parents have been added to create a single variable that summarizes family size. Titles needed simplifying: 'Mlle' and 'Ms' have been recoded as 'Miss', 'Mme' as 'Mrs', and a range of other titles are all coded as 'Rare titles'.*

It should be clear that, apart from the coding skills required, considerable judgement and background knowledge may be needed in simply getting the data ready for analysis, for example using any available cabin information to determine position on the ship. No doubt I could have done this better.

Figure 6.2 shows the proportion of different categories of passenger that survived, for the 897 passengers in the training set. All of these features have predictive ability on their own, with higher survival rates among passengers who are travelling in a better class of the ship, are female, children, paid more for their ticket, had a moderate size family, and had the title

---

* These include Dona, Lady, Countess, Capt, Col, Don, Dr, Major, Rev., Sir, Jonkheer.

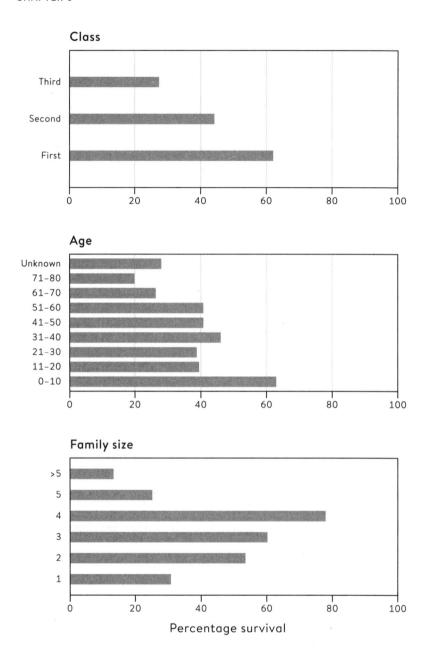

Figure 6.2
Summary survival statistics for training set of 897 *Titanic* passengers, showing the percentage of different categories that survived.

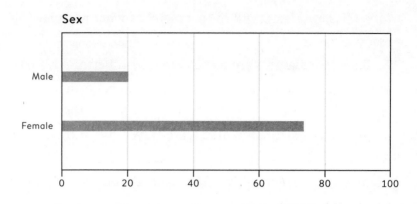

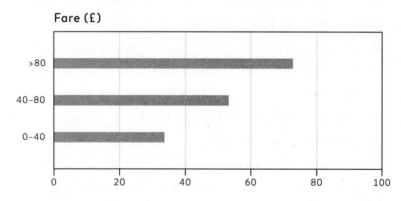

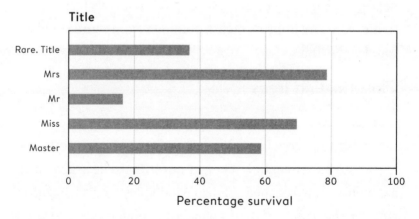

Percentage survival

Mrs, Miss, or Master. All of this matches what we might already suspect.

But these features are not independent. Better-class passengers presumably paid more for their tickets, and may be expected to be travelling with fewer children than would poorer emigrants. Many men were travelling on their own. And the specific coding may be important: should age be considered as a categorical variable, banded into the categories shown in Figure 6–2, or a continuous variable? Competitors have spent a lot of time looking at these features in detail and coding them up to extract the maximum information, but we shall instead proceed straight to making predictions.

Suppose we made the (demonstrably incorrect) prediction that 'Nobody survived'. Then, since 61% of the passengers died, we would get 61% right in the training set. If we used the slightly more complex prediction rule, 'All women survive and no men survive', we would correctly classify 78% of the training set. These naïve rules serve as good baselines from which to measure any improvements obtained from more sophisticated algorithms.

Classification Trees

A **classification tree** is perhaps the simplest form of algorithm, since it consists of a series of yes/no questions, the answer to each deciding the next question to be asked, until a conclusion is reached. Figure 6.3 displays a classification tree for the *Titanic* data, in which passengers are allocated to the majority outcome at the end of the branch. It is easy to see the factors that have been chosen, and the final conclusion. For example, Francis Somerton was titled 'Mr' in the

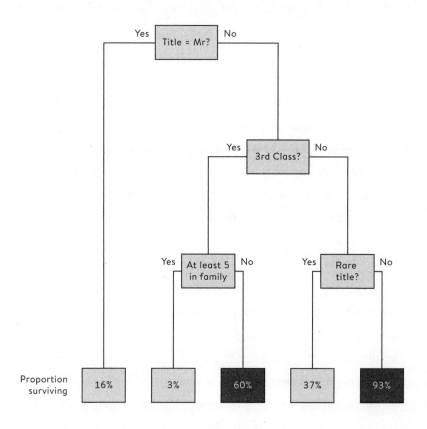

Figure 6.3
A classification tree for the *Titanic* data in which a sequence of
questions leads a passenger to the end of a branch, at which point
they are predicted to survive if the proportion of similar people in
the training set who survived is greater than 50%; these surviving
proportions are shown at the bottom of the tree. The only people
predicted to survive are third-class women and children from
smaller families, and all women and children in first and second
class, provided they do not have rare titles.

database, and so would take the first left-hand branch. The end of this branch contains 58% of the training set, of which 16% survive. We could therefore assess, based on limited information, that Somerton had a 16% chance of surviving. Our simple algorithm identifies two groups with more than 50% survivors: women and children in first and second classes (as long they do not have a rare title), 93% of whom survive. And women and children from third class, provided they come from large families, in which case 60% survive.

Before seeing how such a tree is actually constructed, we need to decide what performance measures to use in our competition.

Assessing the Performance of an Algorithm

If algorithms are going to compete to be the most accurate, someone has to decide what 'accurate' means. In Kaggle's *Titanic* challenge this is simply the percentage of passengers in the test set that are correctly classified, and so after competitors build their algorithm, they upload their predictions for the response variable in the test set and Kaggle measures their accuracy.* We will present results for the whole test set at once (emphasizing that this is not the same as the Kaggle test set).

The classification tree shown in Figure 6.3 has an accuracy of 82% when applied to the training data on which it was

---

* In order not to have to wait until the end of the competition (in 2020 for the *Titanic* data) before anyone gets any feedback, Kaggle splits the test set into public and private sets. Competitors' accuracy scores on the public set are published on a leader board, and this provides a provisional ranking for all to see. But the performance on the private set is what is actually used to evaluate the final ranking of the competitors when the competition closes.

developed. When the algorithm is applied to the test set the accuracy drops slightly to 81%. The numbers of the different types of errors made by the algorithm are shown in Table 6.1 – this is termed the **error matrix**, or sometimes the confusion matrix. If we are trying to detect survivors, the percentage of true survivors that are correctly predicted is known as the **sensitivity** of the algorithm, while the percentage of true non-survivors that are correctly predicted is known as the **specificity**. These terms arise from medical diagnostic testing.

Although the overall accuracy is simple to express, it is a very crude measure of performance and takes no account of the confidence with which a prediction is made. If we look at the tips of the branches of the classification tree, we can see that the discrimination of the training data is not perfect, and at all branches there are some who survive and some not. The crude allocation rule simply chooses the outcome in the majority, but instead we could assign to new cases a *probability* of surviving corresponding to the proportion in the training set. For example, someone with the title 'Mr' could be given a probability of 16% of surviving, rather than a simple categorical prediction that they will not survive.

Algorithms that give a probability (or any number) rather than a simple classification are often compared using **Receiver Operating Characteristic (ROC) curves**, which were originally developed in the Second World War to analyse radar signals. The crucial insight is that we can vary the threshold at which people are predicted to survive. Table 6.1 shows the effect of using a threshold of 50% to predict someone a 'survivor', giving a specificity and sensitivity in the training set of 0.84 and 0.78 respectively. But we could have demanded a higher probability

	TRAINING SET				TEST SET		
	Predicted not to survive	Predicted to survive			Predicted not to survive	Predicted to survive	
Did not survive	475	93	568		228	45	273
Survived	71	258	329		35	104	139
	546	351	897		263	149	412

Accuracy
$= (475 + 258)/897 = 82\%$

Sensitivity
$= 258/329 = 78\%$

Specificity
$= 475/568 = 84\%$

Accuracy
$= (228 + 104)/412 = 81\%$

Sensitivity
$= 104/139 = 75\%$

Specificity
$= 228/273 = 84\%$

Table 6.1
Error matrix of classification tree on training and test data, showing accuracy (% correctly classified), sensitivity (% of survivors correctly classified) and specificity (% of non-survivors correctly classified).

in order to predict someone survives, say 70%, in which case the specificity and sensitivity would have been 0.98 and 0.50 respectively – with this more stringent threshold, we only identify half the true survivors but make very few false claims of surviving. By considering all possible thresholds for predicting a survivor, the possible values for the specificity and sensitivity form a curve. Note that the specificity axis conventionally decreases from 1 to 0 when drawing an ROC curve.

Figure 6.4 shows the ROC curves for training and test sets. A completely useless algorithm that assigns numbers at random would have a diagonal ROC curve, whereas the best algorithms will have ROC curves that move towards the top-left corner. A standard way of comparing ROC curves is by measuring the area underneath them, right down to the horizontal – this will be 0.5 for a useless algorithm, and 1 for a perfect one that gets everyone right. For our *Titanic* test set data, the area under the ROC curve is 0.82. It turns out that there is an elegant interpretation of this area: if we pick a true survivor and a true non-survivor at random, there is an 82% chance that the algorithm gives the true survivor a higher probability of surviving than the true non-survivor. Areas above 0.8 represent fairly good discriminatory ability.

The area under the ROC curve is one way of measuring how well an algorithm splits the survivors from the non-survivors, but it does not measure how good the probabilities are. And the people who are most familiar with probabilistic predictions are weather forecasters.

Suppose we want to predict whether or not it will rain tomorrow at a particular time and place. Basic algorithms

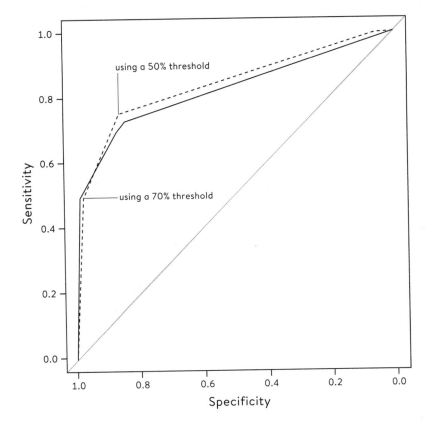

Figure 6.4
ROC curves for the classification tree of Figure 6.3 applied to
training (dashed line) and test (solid line) sets. 'Sensitivity' is the
proportion of survivors correctly identified. 'Specificity' is the
proportion of non-survivors correctly labelled as not surviving.
Areas under curves are 0.84 and 0.82 for training and test sets
respectively.

How do we know how good 'probability of precipitation' forecasts are?

might simply produce a yes/no answer, which might end up being right or wrong. More sophisticated models might produce a probability of it raining, which allows more fine-tuned judgements – the action you take if the algorithm says a 50% chance of rain might be rather different than if it says 5%.

In practice weather forecasts are based on extremely complex computer models which encapsulate detailed mathematical formulae representing how weather develops from current conditions, and each run of the model produces a deterministic yes/no prediction of rain at a particular place and time. So to produce a **probabilistic forecast**, the model has to be run many times starting at slightly adjusted initial conditions, which produces a list of different 'possible futures', in some of which it rains and in some it doesn't. Forecasters run an 'ensemble' of, say, fifty models, and if it rains in five of those possible futures in a particular place and time, they claim a 'probability of precipitation' of 10%.

But how do we check how good these probabilities are? We cannot create a simple error matrix as in the classification tree, since the algorithm is never declaring categorically whether it will rain or not. We can create ROC curves, but these only examine whether days when it rains get higher predictions than when it doesn't. The critical insight is that we also need **calibration**, in the sense that if we take all the days in which the forecaster says 70% chance of rain, then it really

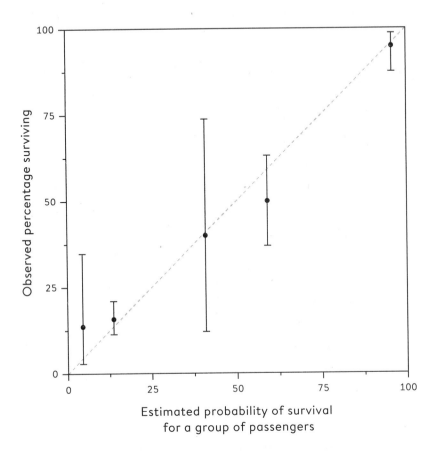

Figure 6.5
Calibration plot for the simple classification tree that provides probabilities of surviving the *Titanic* sinking, in which the observed proportion of survivors on the y-axis is plotted against the predicted proportion on the x-axis. We want the points to lie on the diagonal, showing the probabilities are reliable and mean what they say.

should rain on around 70% of those days. This is taken very seriously by weather forecasters – probabilities should mean what they say, and not be either over- or under-confident.

Calibration plots allow us to see how reliable the stated probabilities are, by collecting together, say, the events given a particular probability of occurrence, and calculating the proportion of such events that actually occurred.

Figure 6.5 shows the calibration plot for the simple classification tree applied to the test set. We want the points to lie near the diagonal since that is where the predicted probabilities match the observed percentages. The vertical bars indicate a region in which we would, given reliable predicted probabilities, expect the actual proportion to lie in 95% of cases. If these include the diagonal line, as in Figure 6.5, we can consider our algorithm to be well calibrated.

A Combined Measure of 'Accuracy' for Probabilities

While the ROC curve assesses how well the algorithm splits the groups, and the calibration plot checks whether the probabilities mean what they say, it would be best to find a simple composite measure that combines both aspects into a single number we could use to compare algorithms. Fortunately weather forecasters back in the 1950s worked out exactly how to do this.

If we were predicting a numerical quantity, such as the temperature at noon tomorrow in a particular place, the accuracy would usually be summarized by the error – the difference between the observed and predicted temperature. The usual summary of the error over a number of days is the **mean-squared-error (MSE)** – this is the average of the squares of the errors, and is analogous to the least-squares criterion we saw used in regression analysis.

The trick for probabilities is to use the same mean-squared-error criterion as when predicting a quantity, but treating a future observation of 'rain' as taking on the value 1, and 'no rain' as being 0. Table 6.2 shows how this would work for a fictitious forecasting system. On Monday a probability of 0.1 is given to rain, but it turns out to be dry (true response is 0), and so the error is $0 - 0.1 = - 0.1$. This is squared to give 0.01, and so on across the week. Then the average of these squared errors, $B = 0.11$, is a measure of the forecaster's (lack of) accuracy.* The average mean-squared-error is known as the **Brier score**, after meteorologist Glenn Brier, who described the method in 1950.

Unfortunately the Brier score is not easy to interpret on its own, and so it is difficult to get a feeling of whether any forecaster is doing well or badly; it is therefore best to compare it with a reference score derived from historical climate records. These 'climate-based' forecasts take no notice whatever of current conditions and simply state the probability of precipitation as the proportion of times in climate history in which it rained on this day. Anyone can make this forecast without any skill whatsoever – in Table 6.2 we assume this means quoting a 20% probability of rain for every day that week. This gives a Brier score for climate (which we call BC) of 0.28.

---

* It might be tempting to use the 'absolute error', meaning you would lose 0.1 when giving a 10% probability to an event that does not happen, as opposed to the squared error of 0.01. This apparently innocuous choice would be a big, big mistake. Some fairly basic theory shows that this 'absolute' penalty would lead people to rationally exaggerate their confidence in order to minimize their expected error, and state '0%' chance of rain, even if they genuinely thought the probability was 10%.

	Monday	Tuesday	Wednesday	Thursday	Friday	Mean-squared-error (Brier Score)
'Probability of precipitation'	0.1	0.2	0.5	0.6	0.3	
Did it actually rain?	No	No	Yes	Yes	No	
True response	0	0	1	1	0	
Error	−0.1	−0.2	0.5	0.4	−0.3	
Squared error	0.01	0.04	0.25	0.16	0.09	B = 0.54/5 = 0.11
Probability from climate	0.2	0.2	0.2	0.2	0.2	
Climate error	−0.2	−0.2	0.8	0.8	0.2	
Squared climate error	0.04	0.04	0.64	0.64	0.04	BC = 1.4/5 = 0.28

Table 6.2
Fictional 'probability of precipitation' forecasts of whether it will rain or not at midday next day at a specific location, with the observed outcome: 1 = did rain, 0 = did not rain. The 'error' is the difference between the predicted and observed outcome, and the mean-squared-error is the Brier score (B). The climate Brier score (BC) is based on using simple long-term average proportions of rain at this time of year as probabilistic forecasts, in this case assumed to be 20% for all days.

Any decent forecasting algorithm should perform better than predictions based on climate alone, and our forecast system has improved the score by BC – B = 0.28 – 0.11 = 0.17. Forecasters then create a 'skill score', which is the proportional reduction of the reference score: in our case, 0.61,[*] meaning our algorithm has made a 61% improvement on a naïve forecaster who uses only climate data.

Clearly our target is 100% skill, but we would only get this if our observed Brier score is reduced to 0, which only happens if we exactly predict whether it will rain or not. This is expecting rather a lot of any forecaster, and in fact skill scores for rain forecasting are now around 0.4 for the following day, and 0.2 for forecasting a week in the future.[2] Of course the laziest prediction is simply to say that whatever happened today will also happen tomorrow, which provides a perfect fit to historical data (today), but may not do particularly well in predicting the future.

When it comes to the *Titanic* challenge, consider the naïve algorithm of just giving everyone a 39% probability of surviving, which is the overall proportion of survivors in the training set. This does not use any individual data, and is essentially the equivalent to predicting weather using climate records rather than information on the current circumstances. The Brier score for this 'skill-less' rule is 0.232.

In contrast, the Brier score for the simple classification tree is 0.139, which is a 40% reduction from the naïve prediction, and so demonstrates considerable skill. Another way of

---

[*] The skill score is (BC–B)/BC = 1 – B/BC = 1 – 0.11/0.28 = 0.61

interpreting this Brier score of 0.139 is that it is exactly what would be obtained had you given all survivors a 63% chance of surviving, and all non-survivors a 63% chance of not surviving.

We shall see if we can improve on this score with some more complicated models, but first we need to issue a warning that they should not get *too* complicated.

Over-fitting

We do not need to stop at the simple classification tree shown in Figure 6.3. We could go on making the tree more and more complex by adding new branches, and this will allow us to correctly classify more of the training set as we identify more and more of its idiosyncrasies.

Figure 6.6 shows such a tree, grown to include many detailed factors. This has an accuracy on the training set of 83%, better than the smaller tree. But when we apply this algorithm to the test data its accuracy drops to 81%, the same as the small tree, and its Brier score is 0.150, clearly worse than the simple tree's 0.139. We have adapted the tree to the training data to such a degree that its predictive ability has started to decline.

This is known as **over-fitting**, and is one of the most vital topics in algorithm construction. By making an algorithm too complex, we essentially start fitting the noise rather than the signal. Randall Munroe (the cartoonist known for his *xkcd* comic strip) produced a fine illustration of over-fitting, by finding plausible 'rules' that US Presidents had followed, only for each to be broken at subsequent elections.[3] For example,

- 'No Republican has won without winning the House or Senate' – until Eisenhower did in 1952.

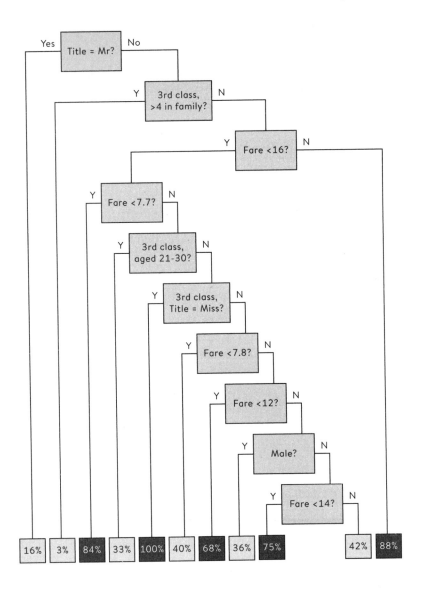

Figure 6.6
Over-fitted classification tree for the *Titanic* data. As in Figure 6.3, the percentage at the end of each branch is the proportion of passengers in the training set who survived, and a new passenger is predicted to survive if this percentage is greater than 50%. The rather strange set of questions suggests the tree has adapted too much to individual cases in the training set.

- 'Catholics can't win' – until Kennedy in 1960.
- 'No one has been elected President after a divorce' – until Reagan in 1980.

and so on, including some clearly over-refined rules such as

- 'No Democratic incumbent without combat experience has beaten someone whose first name is worth more in Scrabble' – until Bill (6 Scrabble points) Clinton beat Bob (7 Scrabble points) Dole in 1996.

We over-fit when we go too far in adapting to local circumstances, in a worthy but misguided effort to be 'unbiased' and take into account all the available information. Usually we would applaud the aim of being unbiased, but this refinement means we have less data to work on, and so the reliability goes down. Over-fitting therefore leads to less bias but at a cost of more uncertainty or variation in the estimates, which is why protection against over-fitting is sometimes known as the **bias/variance trade-off**.

We can illustrate this subtle idea by imagining a huge database of people's lives that is to be used to predict your future health – say your chance of reaching the age of eighty. We could, perhaps, look at people of your current age and socio-economic status, and see what happened to them – there might be 10,000 of these, and if 8,000 reached eighty, we might estimate an 80% chance of people like you reaching eighty, and be very confident in that number since it is based on a lot of people.

But this assessment only uses a couple of features to match you to cases in the database, and ignores more individual characteristics that might refine our prediction – for example no attention is paid to your current health or your habits. A different strategy would be to find people who matched you much more closely, with the same weight, height, blood pressure, cholesterol, exercise, smoking, drinking, and so on and on: let's say we kept on matching on more and more of your personal characteristics until we narrowed it down to just two people in the database who were an almost perfect match. Suppose one had reached eighty and one had not. Would we then estimate a 50% chance of you reaching 80? That 50% figure is in a sense less biased, as it matches you so closely, but, because it is only based on two people, it is not a reliable estimate (i.e., it has large variance).

Intuitively we feel that there is a happy medium between these two extremes; finding that balance is tricky, but crucial. Techniques for avoiding over-fitting include regularization, in which complex models are encouraged but the effects of the variables are pulled in towards zero. But perhaps the most common protection is to use the simple but powerful idea of **cross-validation** when constructing the algorithm.

It is essential to test any predictions on an independent test set that was not used in the training of the algorithm, but that only happens at the end of the development process. So although it might show up our over-fitting at that time, it does not build us a better algorithm. We can, however, mimic having an independent test set by removing say 10% of the training data, developing the algorithm on the remaining 90%, and testing on the removed 10%. This is cross-validation, and can be carried out systematically by removing 10% in turn and repeating the procedure ten times, a procedure known as tenfold cross-validation.

All the algorithms in this chapter have some tunable parameters which are mainly intended to control the complexity of the final algorithm. For example, the standard procedure for building classification trees is to first construct a very deep tree with many branches that is deliberately overfitted, and then prune the tree back to something simpler and more robust: this pruning is controlled by a complexity parameter.

This complexity parameter can be chosen by the cross-validation process. For each of the ten cross-validation samples, a tree is developed for each of a range of different complexity parameters. For each value of the parameter, the average predictive performance over all the ten cross-validation test sets is calculated – this average performance will tend to improve up to a certain point, and then get worse as the trees become too complex. The optimal value for the complexity parameter is the one that gives the best cross-validatory performance, and this value is then used to construct a tree from the complete training set, which is the final version.

Tenfold cross-validation was used to select the complexity parameter in the tree in Figure 6.3, and to choose tuning parameters in all the models we consider below.

Regression Models

We saw in Chapter 5 that the idea of a regression model is to construct a simple formula to predict an outcome. The response variable in the *Titanic* data is a yes/no outcome indicating survival or not, and so a logistic regression is appropriate, just as for the child heart surgery data in Figure 5.2.

Table 6.3 shows the results from fitting a logistic regression. This has been trained using 'boosting', an iterative procedure designed to pay more attention to more difficult cases: individuals in the training set that are incorrectly classified at one iteration are given greater weight in the next iteration, with the number of iterations chosen using tenfold cross-validation.

The coefficients for the features of a particular passenger can be added up to give a total survival score. For example Francis Somerton would start with 3.20, subtract 2.30 for being in third class and 3.86 for being titled 'Mr', but then have 1.43 added back on for being a male in third class. He loses 0.38 for being in a family of one, giving a total score of – 1.91, which translates to a probability of 13% of surviving, slightly less than the 16% given by the simple classification tree.*

This is a 'linear' system, but note that **interactions** have been included which are essentially more complex, combined features, for example the positive score for the interaction of being in third class *and* a male helps counteract the extreme negative scores for the third class and 'Mr' already taken into account. Although we are focusing on predictive performance, these coefficients do provide some interpretation of the importance of different features.

Many more sophisticated regression approaches are available for dealing with large and complex problems, such as non-linear models and a process known as the LASSO, that simultaneously estimates coefficients and selects relevant

---

* To transform a total score S to a survival probability p, use the formula $p = 1/(1 + e^{-S})$, where e is the exponential constant. This is the inverse of the logistic regression equation $\log_e p/(1-p) = S$.

Characteristic	Score
Starting score	3.20
Third class	−2.30
'Mr'	−3.86
Male in third class	+1.43
Rare Title	−2.73
Aged 51–60 in second class	−3.62
Each member of family	−0.38

Table 6.3
Coefficients applied to features in logistic regression for *Titanic* survivor data: negative coefficients decrease the chance of surviving, positive coefficients increase the chance.

predictor variables, essentially by estimating their coefficients to be zero.

More Complex Techniques

Classification trees and regression models arise from somewhat different modelling philosophies: trees attempt to construct simple rules that identify groups of cases with similar expected outcomes, while regression models focus on the weight to be given to specific features, regardless of what else is observed on a case.

The machine learning community makes use of classification trees and regressions, but has developed a wide range of alternative, more complex methods for developing algorithms. For example:

- *Random forests* comprise a large number of trees, each producing a classification, with the final classification decided by a majority vote, a process known as bagging.
- *Support vector machines* try to find linear combinations of features that best split the different outcomes.
- *Neural networks* comprise layers of nodes, each node depending on the previous layer by weights, rather like a series of logistic regressions piled on top of each other. Weights are learned by an optimization procedure, and, rather like random forests, multiple neural networks can be constructed and averaged. Neural networks with many layers have become known as deep-learning models: Google's Inception image-recognition system is said to have over twenty layers and over 300,000 parameters to estimate.

- *K-nearest-neighbour* classifies according to the majority outcome among close cases in the training set.

The results of applying some of these methods to the *Titanic* data, with tuning parameters chosen using tenfold cross-validation and ROC as an optimization criterion, are shown in Table 6.4.

The high accuracy of the naïve rule, 'All females survive, all males do not', which either beats or is close behind more complex algorithms, demonstrates the inadequacy of crude 'accuracy' as a measure of performance. The random forest produces the best discrimination reflected in the area under the ROC curve, although perhaps surprisingly the probabilities coming from the simple classification tree have the best Brier score. There is therefore no clear winning algorithm. Later, in Chapter 10, we shall check whether we can confidently claim there is a proper winner on any of these criteria, since the winning margins might be so small that it can be explained by chance variation – say in who happened to end up in the test and training set.

This reflects a general concern that algorithms that win Kaggle competitions tend to be very complex in order to achieve that tiny final margin needed to win. A major problem is that these algorithms tend to be inscrutable black boxes – they come up with a prediction, but it is almost impossible to work out what is going on inside. This has three negative aspects. First, extreme complexity makes implementation and upgrading a great effort: when Netflix offered a $1m prize for prediction recommendation systems, the winner was so complicated that Netflix ended up not using it. The second

Method	Accuracy (high is good)	Area under ROC curve (high is good)	Brier score (low is good)
Everyone has a 39% chance of surviving	0.639	0.500	0.232
All females survive, all males do not	0.786	0.578	0.214
Simple classification tree	**0.806**	0.819	**0.139**
Classification tree (over-fitted)	**0.806**	0.810	0.150
Logistic regression	0.789	0.824	0.146
Random forest	0.799	**0.850**	0.148
Support Vector Machine (SVM)	0.782	0.825	0.153
Neural network	0.794	0.828	0.146
Averaged neural network	0.794	0.837	0.142
K-nearest-neighbour	0.774	0.812	0.180

Table 6.4
The performance of different algorithms on *Titanic* test data: bold indicates the best results. Complex algorithms have been optimized to maximize the area under the ROC curve.

negative feature is that we do not know how the conclusion was arrived at, or what confidence we should have in it: we just have to take it or leave it. Simpler algorithms can better explain themselves. Finally, if we do not know how an algorithm is producing its answer, we cannot investigate it for implicit but systematic biases against some members of the community – a point I expand on below.

All this points to the possibility that quantitative performance may not be the sole criterion for an algorithm, and once performance is 'good enough', it may be reasonable to trade off further small increases for the need to retain simplicity.

Who was the luckiest person on the *Titanic*?

The survivor with the highest Brier score when averaged over all the algorithms might be considered the most surprising survivor. This was Karl Dahl, a 45-year-old Norwegian/Australian joiner travelling on his own in third class, who had paid the same fare as Francis Somerton; two algorithms even gave him a 0% chance of surviving. He apparently dived into the freezing water and clambered into Lifeboat 15, in spite of some on the lifeboat trying to push him back. Maybe he just used his strength.

This is in stark contrast to Francis Somerton from Ilfracombe, whose death, we have found, fitted into the general pattern. Rather than having a successful husband in America, his wife Hannah Somerton was left just £5, less than Francis spent on his ticket.

Challenges of Algorithms

Algorithms can display remarkable performance, but as their role in society increases so their potential problems become highlighted. Four main concerns can be identified.

- *Lack of robustness*: Algorithms are derived from associations, and since they do not understand underlying processes, they can be overly sensitive to changes. Even if we are only concerned with accuracy rather than scientific truth, we still need to remember the basic principles of the PPDAC cycle, and the stages of going from the data obtained from a sample through to statements being made about a target population. For predictive analytics, this target population comprises future cases, and if everything stays the same, then algorithms constructed on past data should perform well. But the world does not always stay the same. We've noted the failure of algorithms in the changing financial world of 2007–8, and another notable example was the attempt by Google to predict flu trends based on the pattern of search terms being submitted by users. This initially performed well but then in 2013 started to dramatically over-predict flu rates: one explanation is that changes introduced by Google into the search engine may have led to more search terms that pointed to flu.
- *Not accounting for statistical variability*: Automated rankings based on limited data will be unreliable. Teachers in the US have been ranked and penalized for the performance of their students in a single year,

although class sizes of less than thirty do not provide a reliable basis for assessing the value added by a teacher. This will reveal itself in teachers having implausibly dramatic changes in annual assessment: in Virginia, a quarter of teachers showed more than 40-point differences in a 1–100 scale from year-to-year.[*]

- *Implicit bias*: To repeat, algorithms are based on associations, which may mean they end up using features that we would normally think are irrelevant to the task in hand. When a vision algorithm was trained to discriminate pictures of huskies from German Shepherds, it was very effective until it failed on huskies that were kept as pets – it turned out that its apparent skill was based on identifying snow in the background.[4] Less trivial examples include an algorithm for identifying beauty that did not like dark skin, and another that identified Black people as gorillas. Algorithms that can have a major impact on people's lives, such as those deciding credit ratings or insurance, may be banned from using race as a predictor but might use postcodes to reveal neighbourhood, which is a strong proxy for race.

- *Lack of transparency*: Some algorithms may be opaque due to their sheer complexity. But even simple regression-based algorithms become totally inscrutable if their structure is private, perhaps through being a proprietary commercial product. This is one of the major complaints about so-called recidivism algorithms, such

---

[*] From Cathy O'Neil's book *Weapons of Math Destruction*, which provides many examples of the misuse of algorithms.

as Northpointe's Correctional Offender Management Profiling for Alternative Sanctions (COMPAS) or MMR's Level of Service Inventory – Revised (LSI-R).[5] These algorithms produce a risk-score or category that can be used to guide probation decisions and sentencing, and yet the way in which the factors are weighted is unknown. Furthermore, since information about upbringing and past criminal associates is collected, decisions are not solely based on a personal criminal history but background factors that have been shown to be associated with future criminality, even if the underlying common factor is poverty and deprivation. Of course, if all that mattered was accurate prediction, then anything goes and any factor, even including race, might be used. But many argue that fairness and justice demand that these algorithms should be controlled, transparent and able to be appealed against.

Even for proprietary algorithms, some degree of explanation is possible provided we can experiment with different inputs. When purchasing online insurance, the quoted premium is calculated according to an unknown formula subject only to certain legal constraints: for example, in the UK car insurance quotes cannot take into account the gender of the applicant, life insurance cannot use race or any genetic information except Huntingdon's disease, and so on. But we can still get an idea of the influence of different factors by systematically lying and seeing how the quotation changes: this allows a certain degree of reverse-engineering of the algorithm to see what is driving the premium.

There is increasing demand for accountability of algorithms that affect people's lives, and requirements for comprehensible explanation of conclusions are being built into legislation. These demands militate against complex black boxes, and may lead to a preference for (rather old-fashioned) regression-based algorithms that make the influence of each item of evidence clear.

But having looked at the dark side of algorithms, it is fitting to end with an example that seems entirely beneficial and empowering.

> What is the expected benefit of adjuvant therapy following breast cancer surgery?

Nearly all women newly diagnosed with breast cancer will receive some form of surgery, although this might be limited in extent. A critical issue is then the choice of adjuvant therapy that follows surgery in order to reduce the chances of recurrence and subsequent death from breast cancer, and treatment options may include radiotherapy, hormone therapy, chemotherapy and other drug options. Within the PPDAC cycle, this is the Problem.

The Plan adopted by UK researchers was to develop an algorithm to help with this decision, using Data on 5,700 historical cases of women with breast cancer obtained from the UK Cancer Registry. The Analysis comprised the construction of an algorithm that would use detailed information on the woman and her tumour in order to calculate her chances of survival for up to ten years following surgery, and how these changed with different treatments. But care is required in analysing the outcomes of women given these treatments in the past: they were given the treatments for unknown reasons and we cannot use the apparent benefits observed in the database. Instead a regression model is fitted, with survival as the outcome, but forcing the effect of treatments to be those estimated from reviews of large-scale clinical trials. The subsequent algorithm is publicly available, and its discrimination and calibration has been checked on independent data sets comprising 27,000 women.[6]

The resulting computer software is called Predict 2.1, and the results are Communicated through the proportions of similar women expected to survive five and ten years for different adjuvant treatments. Some results for a fictitious woman are shown in Table 6.5.

Predict 2.1 is not perfect, and the figures in Table 6.5 can only be used as ballpark guides for an individual: they are what we would expect to happen to women who match the features included in the algorithm, and additional factors should be taken into account for a specific woman. Nevertheless, Predict 2.1 is used routinely for tens of thousands of cases a month, both in multidisciplinary team meetings (MDTs) in which a patient's treatment options are formulated, and in communicating that information to the woman. For those women who wish to fully engage in their treatment choices, a process known as 'shared-care', it can provide information normally only available to the clinicians, and empower them to have greater control over their lives. The algorithm is not proprietary, the software is open source and the system is regularly being upgraded

Treatment	Additional benefit over previous treatments	Overall survival %
Surgery only	–	64%
+ Hormone therapy	7%	70%
+ Chemotherapy	6%	76%
+ Trastuzamab (Herceptin)	3%	79%
For women free from cancer		87%

Table 6.5
Using the Predict 2.1 algorithm, the proportion of 65-year-old women expected to survive ten years after surgery for breast cancer, when a 2cm grade 2 tumour was detected at screening, with two positive nodes, and ER, HER2 and Ki-67 status are all positive. The cumulative expected benefits for different adjuvant treatments are shown, although these treatments may have adverse effects. The surviving proportion for 'women free from cancer' represents the best survival achievable, given the age of the woman.

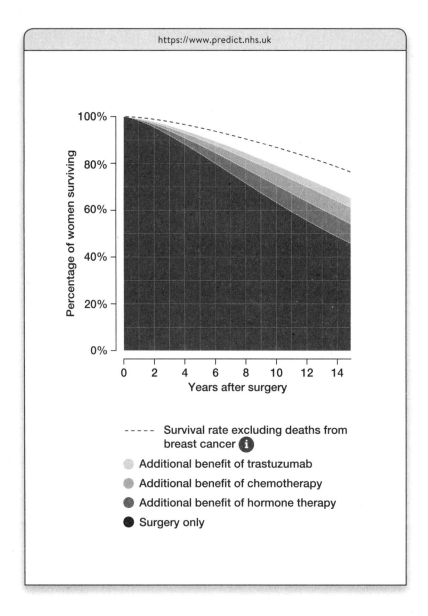

Figure 6.7
Survival curves from Predict 2.1 for up to fifteen years post-surgery, for women with the features listed in the legend to Table 6.5, showing the cumulative additional survival from further treatments. The area above the dashed line represents women with breast cancer who die of other causes.

to provide further information, including adverse effects of treatments.

Artificial Intelligence

Ever since its first use in the 1950s, the idea of artificial intelligence (AI) has received periodic hype and enthusiasm and subsequent troughs of criticism. I was working on computer-aided diagnosis and handling uncertainty in AI in the 1980s, when much of the discourse was framed in terms of a competition between approaches based on probability and statistics, those based on encapsulating expert 'rules' of judgement or those trying to emulate cognitive capacities through neural networks. The field has now matured, with a more pragmatic and ecumenical approach to its underlying philosophy, although the hype has not gone away.

AI comprises intelligence demonstrated by machines, which is a suitably wide-ranging idea. It is a much bigger topic than the restricted issue of algorithms discussed in this chapter, and statistical analysis is only one component to building AI systems. But, as demonstrated by the extraordinary recent achievements of algorithms in vision, speech, games and so on, statistical learning plays a major part in the successes in 'narrow' AI. Systems such as Predict, which previously would be thought of as statistics-based decision-support systems, might now reasonably be called AI.*

Many of the challenges listed above come down to algorithms only modelling associations, and not having an idea of

---

* Perhaps for no other reason than to attract funding . . .

underlying causal processes. Judea Pearl, who has been large-ly responsible for the increased focus on causal reasoning in AI, argues that these models only allow us to answer questions of the type, 'We have observed X, what do we expect to observe next?' Whereas general AI needs a causal model for how the world actually works, which would allow it to answer human-level questions concerning the effect of interventions ('What if we do X?'), and counterfactuals ('What if we hadn't done X?').

We are a long way from AI having this ability.

This book emphasizes the classic statistical problems of small samples, systematic bias (in the statistical sense) and lack of generalizability to new situations. The list of challenges for algorithms shows that although having masses of data may reduce the concern about sample size, the other problems tend to get worse, and we are faced with the additional problem of explaining the reasoning of an algorithm.

Having bucketloads of data only increases the challenges in producing robust and responsible conclusions. A basic humility when building algorithms is crucial.

Summary

- Algorithms built from data can be used for classification and prediction in technological applications.
- It is important to guard against over-fitting an algorithm to training data, essentially fitting to noise rather than signal.
- Algorithms can be evaluated by the classification accuracy, their ability to discriminate between groups, and their overall predictive accuracy.
- Complex algorithms may lack transparency, and it may be worth trading off some accuracy for comprehension.
- The use of algorithms and artificial intelligence presents many challenges, and insights into both the power and limitations of machine-learning methods is vital.

How Sure Can We Be About What Is Going On? Estimates and Intervals

How many people are unemployed in the UK?

In January 2018 the BBC News website announced that over the three months to the previous November, 'UK unemployment fell by 3,000 to 1.44 million'. The reason for this fall was debated, but nobody questioned whether this figure really was accurate. But careful scrutiny of the UK Office of National Statistics website revealed that the **margin of error** on this total was ± 77,000 – in other words, the true change could have been between a fall of 80,000 and a rise of 74,000. So although journalists and politicians appear to believe this claimed decline of 3,000 was a fixed, immutable tally of the entire country, it was in fact an imprecise estimate based on a survey of around 100,000 people.* Similarly, when the US Bureau of Labor Statistics reported a seasonally adjusted rise in civilian unemployment of 108,000 from December 2017 to January 2018, this was based on a sample of around 60,000

---

* When I once suggested to a group of journalists that this should be clearly stated in their articles, I was met with blank incomprehension.

households and had a margin of error (again rather difficult to find) of ± 300,000.[*1]

Acknowledging uncertainty is important. Anyone can make an estimate, but being able to realistically assess its possible error is a crucial element of statistical science. Even though it does involve some challenging concepts.

Suppose that we have collected some accurate data, perhaps with a well-designed survey, and we want to generalize the findings to our study population. If we have been careful and avoided internal biases, say by having a random sample, then we should expect the summary statistics calculated from the sample to be close to the corresponding values for the study population.

This important point is worth elaborating. In a well-conducted study, we expect our sample mean to be close to the population mean, the sample inter-quartile range to be close to the population inter-quartile range, and so on. We saw the idea of population summaries illustrated with the birth-weight data in Chapter 3, where we called the sample mean a statistic, and the population mean a parameter. In more technical statistical writing, these two figures are generally distinguished by giving them Roman and Greek letters respectively, in a possibly doomed attempt to avoid confusion; for example m often represents a sample mean, while the Greek μ (mu) is a population mean, and s generally represents a sample standard deviation, σ (sigma) a population standard deviation.

---

[*] Changes in unemployment derived from payroll data is based on employer returns and is somewhat more accurate, with a margin of error of around ± 100,000.

Often just the summary statistic is communicated, and this may be enough in some circumstances. For example, we have seen that most people are unaware that unemployment figures for the UK and US are not based on a full count of those officially registered as unemployed, but instead on large surveys. If such a survey finds that 7% of the sample are unemployed, national agencies and the media usually present this value as if it is a simple fact that 7% of the whole population are unemployed, rather than acknowledging that 7% is only an estimate. In more precise terms, they confuse the sample mean with the population mean.

This may not matter if we just want to give a broad picture of what is going on in the country, and the survey is huge and reliable. But suppose, to take a rather extreme illustration, that you hear that only 100 people were asked if they were unemployed, and seven said they were. The estimate would be 7%, but you probably wouldn't think it was very reliable, and you would not be very happy at this value being treated as if it described the whole population. What if the survey size were 1,000? 100,000? With a large enough survey, you may start feeling more comfortable with the fact that a sample estimate is a good enough summary. The sample size should affect your confidence in the estimate, and knowing exactly how much difference it makes is a basic necessity for proper statistical inference.

Numbers of Sexual Partners

Let's revisit the Natsal survey in Chapter 2, in which participants were asked how many sexual partners they had had in their lifetime. In the age band of 35–44 there were 1,100 female and 796 male respondents, so it was a large survey from which the sample summary statistics shown in Table 2–2 were calculated, such as the median number of reported partners being 8 for men and 5 for women. Since we know the survey was based on a proper random-sampling scheme, it is fairly reasonable to assume that the study population matches the target population, which is the adult British population. The crucial question is: how close are these statistics to what we would have found had we been able to ask everyone in the country?

As an illustration of how the accuracy of statistics depends on sample size, we shall pretend for the moment that the men in the survey in fact represent the population in which we are interested. The bottom panel of Figure 7.1 shows the distribution of their responses. For illustration, we then take successive samples of individuals from this 'population' of 796 men, pausing when we reach 10, 50 and 200 men. The data distributions of these samples are shown in Figure 7.1 – it is clear that the smaller samples are 'bumpier', since they are sensitive to single data-points. The summary statistics for the successively larger samples are shown in Table 7.1, showing that the rather high number of partners (mean 21.1) in the first sample of ten individuals gets steadily overwhelmed, as the statistics get closer and closer to those of the whole group of 796 men as the sample size increases.

Let's now go back to the actual problem at hand – what can we say about the mean and median number of partners in the entire study population of men between 35 and 44, based on the actual samples of men shown in Figure 7.1? We could estimate these population parameters by the sample statistics

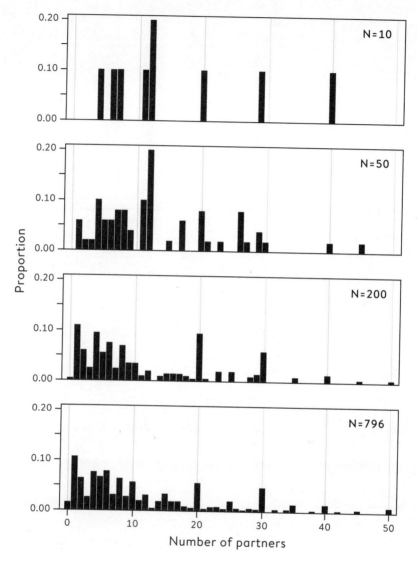

Figure 7.1

The bottom panel shows the distribution of responses of all 796 men in the survey. Individuals are successively sampled at random from this group, pausing at samples of size 10, 50, 200, producing the distributions in the top three panels. Smaller sample sizes show a more variable pattern, but the shape of the distribution gradually approaches that of the whole group of 796 men. Values above 50 partners are not shown.

Size of sample	Mean number of partners	Median number of partners
10	21.1	12
50	18.8	9
200	15.4	8
796	14.9	8

Table 7.1
Summary statistics for the lifetime number of sexual partners reported by men aged 35–44 in Natsal-3, for successively larger random samples and the complete data on 796 men.

of each group shown in Table 7.1, presuming that those based on the bigger samples are somehow 'better': for example the estimates of the mean number of partners are converging towards 15, and with a big enough sample we could presumably get as close as we wanted to the true answer.

Now we come to a critical step. In order to work out how accurate these statistics might be, we need to think of how much our statistics might change if we (in our imagination) were to repeat the sampling process many times. In other words, if we repeatedly drew samples of 796 men from the country, how much would the calculated statistics vary?

If we knew how much these estimates would vary, then it would help tell us how accurate our actual estimate was. But unfortunately we could only work out the precise variability in our estimates if we knew precisely the details of the population. And this is exactly what we do not know.

There are two ways to resolve this circularity. The first is to make some mathematical assumptions about the shape of the population distribution, and use sophisticated probability theory to work out the variability we would expect in our estimate, and hence how far away we might expect, say, the average of our sample to be from the mean of the population. This is the traditional method that is taught in statistics textbooks, and we shall see how this works in Chapter 9.

However, there is an alternative approach, based on the plausible assumption that the population should look roughly like the sample. Since we cannot repeatedly draw a new sample from the population, we instead repeatedly draw new samples from our sample!

We can illustrate this idea with our previous sample of

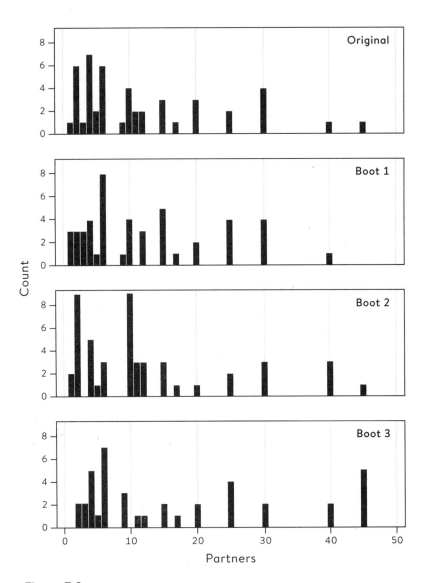

Figure 7.2

The original sample of 50 observations, and three 'bootstrap' resamples, each based on sampling 50 observations at random from the original set, replacing the sampled data-point each time. For example, an observation of 40 partners occurs once in the original sample (on the far right). This data-point was sampled once in the first bootstrap sample, three times in the second, and twice in the third.

50, shown in the top panel of Figure 7.2, which has a mean of 18.8. Suppose we draw 50 data-points in sequence, each time replacing the point we have taken, and get the data distribution shown in the second panel, which has a mean of 14.5. Note that this distribution can only contain data-points taking on the same values as the original sample, but will contain different numbers of each value and so the shape of the distribution will be slightly different, and give a slightly different mean. This can then be repeated, and Figure 7.2 shows three such resamples, with means of 14.5, 26.5 and 22.5.

We therefore get an idea of how our estimate varies through this process of resampling with replacement. This is known as **bootstrapping** the data – the magical idea of pulling oneself up by one's own bootstraps is reflected in this ability to learn about the variability in an estimate without having to make any assumptions about the shape of the population distribution.

If we repeat this resampling, say, 1,000 times, we get 1,000 possible estimates of the mean. These are displayed as histograms in the second panel of Figure 7.3. Other panels show the results of bootstrapping the other samples shown in Figure 7.1, with each histogram showing the spread of bootstrap estimates around the mean of the original sample. These are known as **sampling distributions** of estimates, since they reflect the variability in estimates that arise from repeated sampling of data.

Figure 7.3 displays some clear features. The first, and perhaps most notable, is that almost all trace of the skewness of the original samples has gone – the distributions of the estimates based on the resampled data are almost symmetric

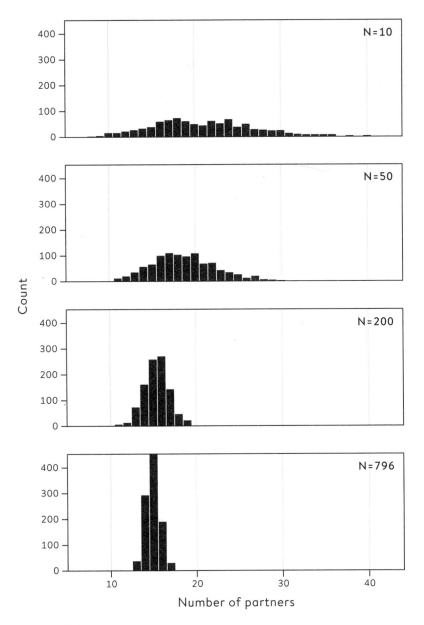

Figure 7.3
Distribution of sample means of 1,000 bootstrap resamples,
for each of the original samples of size 10, 50, 200 and 796 shown
in Figure 7.1. The variability of the sample means of the bootstrap
resamples decreases as the sample size increases.

around the mean of the original data. This is a first glimpse of what is known as the Central Limit Theorem, which says that the distribution of sample means tends towards the form of a normal distribution with increasing sample size, *almost regardless of the shape of the original data distribution.* This is an exceptional result, which we shall explore further in Chapter 9.

Crucially, these bootstrap distributions allow us to quantify our uncertainty about the estimates shown in Table 7.1. For example, we can find the range of values that contains 95% of the means of the bootstrap resamples, and call this a 95% uncertainty interval for the original estimates, or alternatively they can be called margins of error. These are shown in Table 7.2 – the symmetry of the bootstrap distributions means the uncertainty intervals are roughly symmetric around the original estimate.

The second important feature of Figure 7.3 is that the bootstrap distributions get narrower as the sample size increases, which is reflected in the steadily narrower 95% uncertainty intervals.

This section has introduced some difficult but important ideas:

- the variability in statistics based on samples
- bootstrapping data when we do not want to make assumptions about the shape of the population
- the fact that the shape of the distribution of the statistics does not depend on the shape of the original distribution from which the individual data-points are drawn

Size of sample	Mean number of partners	95% bootstrap uncertainty interval
10	21.1	10.4 to 34.6
50	18.8	12.2 to 27.2
200	15.4	12.7 to 18.6
796	14.9	13.4 to 16.6

Table 7.2
Sample means for the lifetime number of sexual partners reported by men aged 35–44 in Natsal-3, for nested random samples of size 10, 50, 200 and complete data on 796 men, with 95% bootstrap uncertainty intervals, also known as margins of error.

Rather remarkably, this has all been accomplished without any mathematics except the idea of drawing observations at random.

I now show that the same bootstrap strategy can be applied to more complex situations.

In Chapter 5 I fitted regression lines to Galton's height data, enabling predictions to be made of, say, a daughter's height based on her mother's height, using a regression line with an estimated gradient of 0.33 (Table 5.2). But how confident can we be about the position of that fitted line? Bootstrapping provides an intuitive way of answering this question without making any mathematical assumptions about the underlying population.

To bootstrap the 433 daughter/mother pairs shown in Figure 7.4, a resample of 433 is drawn, with replacement, from the data, and the least-squares ('best-fit') line fitted. This is repeated as many times as desired: for illustration, Figure 7.4 shows the fitted lines arising from just twenty resamples in order to demonstrate the scatter of lines. It is clear that, since the original data set is large, there is relatively little variability in the fitted lines and, when based on 1,000 bootstrap resamples, a 95% interval for the gradient runs from 0.22 to 0.44.

Bootstrapping provides an intuitive, computer-intensive way of assessing the uncertainty in our estimates, without making strong assumptions and without using probability theory. But the technique is not feasible when it comes to, say, working out the margins of error on unemployment

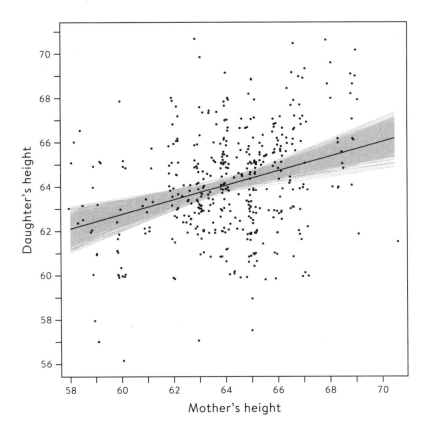

Figure 7.4
Fitted regression lines for twenty bootstrap resamples of Galton's mother–daughter height data superimposed on original data, showing the relatively small variability in gradient due to the large sample size.

surveys of 100,000 people. Although bootstrapping is a simple, brilliant and extraordinarily effective idea, it is just too clumsy to bootstrap such large quantities of data, especially when a convenient theory exists that can generate formulae for the width of uncertainty intervals. But before demonstrating this theory in Chapter 9, we must first face the delightful, but challenging, theory of probability.

Summary

- Uncertainty intervals are an important part of communicating statistics.
- Bootstrapping a sample consists of creating new data sets of the same size by resampling the original data, with replacement.
- Sample statistics calculated from bootstrap re-samples tend towards a normal distribution for larger data sets, regardless of the shape of the original data distribution.
- Uncertainty intervals based on bootstrapping take advantage of modern computer power, do not require assumptions about the mathematical form of the population and do not require complex probability theory.

CHAPTER 8

Probability – the Language of Uncertainty and Variability

In 1650s France, the self-styled Chevalier de Méré had a gambling problem. It was not that he gambled too much (although he did), but he wanted to know which of two games he stood the greatest chance of winning -

> Game 1: Throw a fair die at most four times, and win if you get a six.
>
> Game 2: Throw two fair dice at most twenty-four times, and win if you get a double-six.

Which was his better bet?

Following good empirical statistical principles, the Chevalier de Méré decided to play both games numerous times and see how often he won. This took a great deal of time and effort, but in a bizarre parallel universe in which there were computers but no probability theory, the good Chevalier (real name Antoine Gombaud) would not have wasted his time collecting data on his successes – he would simply have simulated thousands of games.

Figure 8.1 displays the results of such a simulation, showing how the overall proportion of times that he wins each game changes as he 'plays' more and more. Although Game 2 looks the better bet for a while, after around 400 games of

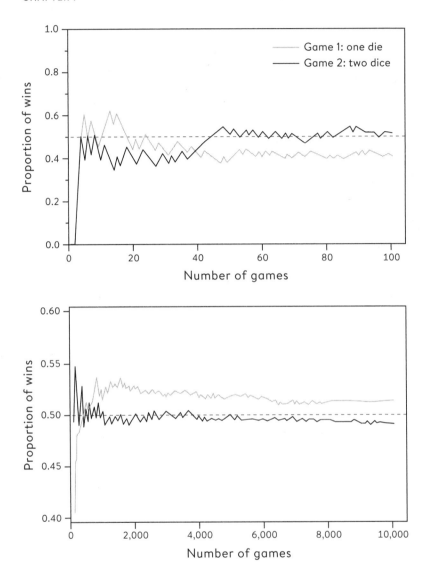

CHAPTER 7

Figure 8.1

A computer simulation of 10,000 repeats of two games. In Game 1, you win if you throw a six in at most 4 throws of a fair die; in Game 2, you win if you throw a double-six in at most 24 throws of two fair dice. Over the first 100 games of each (upper chart) it looks like there is a higher chance of winning Game 2, but after thousands of plays (lower chart) it becomes apparent that Game 1 provides a marginally better bet.

each it becomes clear that Game 1 is better, and in the (very) long run he can expect to win around 52% of Game 1, and only 49% of Game 2.

The Chevalier, rather remarkably, played so often that he came to the same conclusion: Game 1 was the marginally better bet. This went against his (erroneous) attempts at calculating the chances of winning,* and so he turned to the fashionable Mersenne Salon in Paris for help. Fortunately the philosopher Blaise Pascal was also a member of the Salon, and Pascal in turn wrote to his friend Pierre de Fermat (the one with the famous Last Theorem) about the problems presented by the Chevalier. Together they developed the first steps in probability theory.

Despite the fact that for millennia humans have gambled on the way bits of bone or dice would turn up when thrown, the formal theory of probability is a comparatively recent idea. After Pascal and Fermat's work in the 1650s, the mathematical essentials were all sorted in the next fifty years, and now probability has applications in physics, insurance, pensions, financial trading, forecasting, and of course gambling. But why do we need to use probability theory when doing statistics?

We have already seen the concept of data-points being 'picked at random' from a population distribution – the friend

---

* For Game 1, he thought that four throws, with a 1/6 chance each time, would mean an overall chance of winning of 4 times 1/6, which is 2/3. Similarly, for Game 2, he thought that 24 throws with a 1/36 chance each time meant a 24/36 = 2/3 chance, the same as Game 1. These are still common errors made by students – to show that this cannot be right, simply consider the following: if he had 12 throws in Game 1, would the chance of winning be 12 times 1/6 = 2? The correct reasoning is provided in endnote 1.

with the low birth-weight baby in Chapter 3 was our first introduction to probability. We have to assume that anyone in the population is equally likely to be chosen to be part of our sample: remember Gallup's analogy of stirring soup well before tasting. And we have seen that if we want to make statistical inferences about unknown aspects of the world, including making predictions or forecasts, then our conclusions will always have some uncertainty attached to them.

In the last chapter we saw how we could use bootstrapping to see how much variation we would expect our summary statistic to have if we repeated the sampling process again and again, and then used this variability to express our uncertainty about the true, but unknown, characteristics of the population. This again only needs the idea of 'picking at random', an idea that even small children can easily grasp as representing a fair choice.

Traditionally a statistics course would start with probability – that is how I have always begun when teaching in Cambridge – but this rather mathematical initiation can be an obstruction to grasping all the important ideas in the preceding chapters that did not require probability theory. In contrast, this book is part of what could be called a new wave in statistics teaching, in which formal probability theory as a basis for statistical inference does not come in till much later.[2] We have seen that computer simulation is a very powerful tool for both exploring possible future events and bootstrapping historical data, but it is a rather clumsy and brute-force way of carrying out statistical analysis. So although we have got a long way while avoiding formal probability theory, it is time to face up to its vital role in providing 'the language of uncertainty'.

But why the reluctance to use this brilliant theory developed over the last 350 years? I am often asked why people tend to find probability a difficult and unintuitive idea, and I reply that, after forty years researching and teaching in this area, I have finally concluded that it is because probability really *is* a difficult and unintuitive idea. I have sympathy for anyone who finds probability tricky. Even after my decades as a statistician, when asked a basic school question using probability, I have to go away, sit in silence with a pen and paper, try it a few different ways, and finally announce what I hope is the correct answer.

Let's start with my favourite problem-solving technique, which might have saved some politicians some embarrassment.

The Rules of Probability Made, Possibly, a Bit Simpler

In 2012, 97 Members of Parliament in London were asked: 'If you spin a coin twice, what is the probability of getting two heads?' The majority, 60 out of 97, could not give the correct answer.* How might these politicians have done better?

Perhaps they should have known the rules of probability, but most people don't. But an alternative would be to use a more intuitive idea which has been shown in numerous psychology experiments to improve people's reasoning about probability.

This is the idea of 'expected frequency'. When faced with the problem of the two coins, you ask yourself, 'What would I expect to happen if I tried the experiment a number of times?' Let's say that you tried flipping first one coin, and

---

* Spoiler alert: the answer is ¼, or 25%, or 0.25.

then another, a total of four times. I suspect that even a politician could, with a bit of thought, conclude that they would expect to get the results shown in Figure 8.2.

So 1 in 4 times you would expect to get two heads. Therefore, the reasoning goes, the probability that on a particular attempt you would get two heads is 1 in 4, or ¼. Which, fortunately, is the correct answer.

This expected frequency tree can be transformed into a 'probability tree' by labelling each 'split' with the fraction of occasions that it is taken (Figure 8.3). It should then be clear that the overall probability for an entire 'branch' of the tree, say a head followed by a head, is obtained by multiplying the fractions on the splits along the branch, so that ½ × ½ = ¼.

Probability trees are a widespread and extremely effective way of teaching probability in school. Indeed, we can use this simple example of flipping two coins to see all the rules of probability, since the probability tree shows that

1. *The probability of an event is a number between 0 and 1:* 0 for impossible events (for example, flip no heads and no tails), 1 for certain events (flip any of the four possible combinations).

2. *Complement rule:* the probability of an event happening is one minus the probability of it not happening. For example, the probability of 'at least one tail' is one minus the probability of 'two heads': 1 − ¼ = ¾.

3. *The addition, or the OR, rule:* add probabilities of mutually exclusive events (meaning they cannot both happen at the same time) to get the total probability. For example,

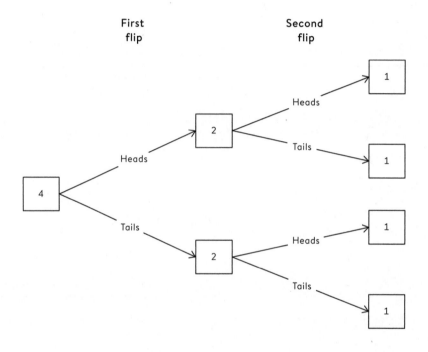

Figure 8.2
Expected frequency tree for two coin flips repeated four times. For example, you expect two out of the four first flips to be heads, and then one each of these to be heads and tails on the second flip.

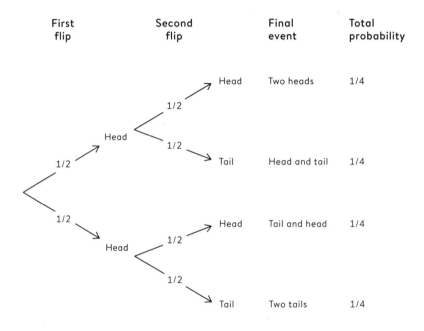

First flip	Second flip		Final event	Total probability
		Head	Two heads	1/4
	1/2			
Head				
	1/2			
1/2		Tail	Head and tail	1/4
		Head	Tail and head	1/4
1/2	1/2			
Head				
	1/2			
		Tail	Two tails	1/4

Figure 8.3
Probability tree for flipping two coins. Each 'split' is labelled with the fraction of occasions that it is taken. The probability for an entire 'branch' of the tree is obtained by multiplying the fractions on the splits along the branch.

the probability of 'at least one head' is ¾, since it comprises 'two heads' OR 'head+tail' OR 'tail+head', each with a probability of ¼.

4. *The multiplication, or the AND, rule*: multiply probabilities to get the overall probability of a sequence of **independent events** (meaning one does not affect the other) occurring. For example, the probability of a head AND a head is $\frac{1}{2} \times \frac{1}{2} = \frac{1}{4}$.

These basic rules allow us to solve the Chevalier de Méré's gambling problem, revealing that he does indeed have a 52% chance of winning Game 1, and a 49% chance of winning Game 2.[1]

We are still making strong assumptions, even in this simple coin-flipping example. We are assuming the coin is fair and balanced, it is flipped properly so the result is not predictable, it does not land on its edge, an asteroid does not strike after the first flip, and so on. These are serious considerations (except possibly the asteroid): they serve to emphasize that all the probabilities we use are *conditional* – there is no such thing as the unconditional probability of an event; there are always assumptions and other factors that could affect the probability. And, as we now see, we need to be careful about what we condition on.

Conditional Probability – When Our Probabilities Depend on Other Events

> When screening for breast cancer, mammography is roughly 90% accurate, in the sense that 90% of women with cancer, and 90% of women without cancer, will be correctly classified. Suppose 1% of women being screened actually have cancer: what is the probability that a randomly chosen woman will have a positive mammogram, and if she does, what is the chance she really has cancer?

In the two-coins case, the events were independent, in that the probability of flipping a head on the second flip did not depend on what the first flip was. In school we usually learn about **dependent events** by being asked somewhat tedious questions about, say, a series of different coloured socks being drawn from a drawer. The example above is somewhat more relevant to real life.

This kind of problem is a classic in intelligence tests and is not an easy problem to solve, but using the idea of expected frequency it becomes remarkably straightforward. The crucial idea is to think of what we would expect to happen to a large group of women, say 1,000, as shown in Figure 8.4.

Out of the 1,000 women, 10 (1%) actually have breast cancer. Of these 10, 9 (90%) have a positive result. But out of the 990 women without cancer, 99 (10%) are falsely given a positive mammography result. Putting these together makes

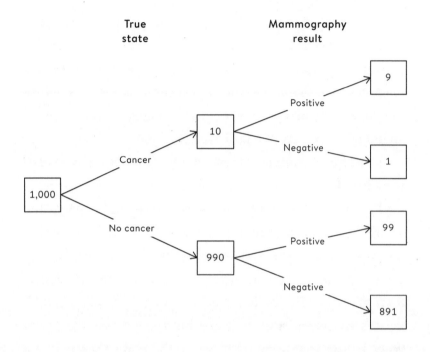

Figure 8.4
Expected frequency tree showing what we would expect to happen to 1,000 women being screened for breast cancer. We assume that 1% of the women have breast cancer, and mammography correctly classifies 90% of women with cancer, and 90% of women without cancer. Overall we would expect 9 + 99 = 108 positive mammograms, of which 9 truly have cancer.

a total of 9 + 99 = 108 positive mammograms, and so the probability that a randomly chosen woman will get a positive result is 108/1,000, or around 11%. But of these 108, only 9 truly have cancer, and so there is only a 9/108 = 8% probability the woman actually has cancer.

This exercise in conditional probability helps us to understand a very counter-intuitive result: in spite of the '90% accuracy' of the scan, the vast majority of women with a positive mammogram do not have breast cancer. It is easy to confuse the probability of a positive test, given cancer, with the probability of cancer, given a positive test.

This type of confusion is popularly known as the **prosecutor's fallacy**, since it is so prevalent in court cases involving DNA. A forensic expert might claim, for example, that 'if the accused is innocent, there is only a 1 in a billion chance that they would match the DNA found at the crime scene'. But this is wrongly interpreted as meaning 'given the DNA evidence, there is only a 1 in a billion chance the accused is innocent'.*

It is an easy mistake to make, but the logic is as faulty as going from the statement 'if you're the Pope, then you're a Catholic' to 'if you're a Catholic, then you're the Pope', where the flaw is somewhat simpler to spot.

What Is 'Probability' Anyway?

In school we are taught about the mathematics of distance, weight and time – which we can measure with a ruler, scales

---

* This is also known as the 'law of the transposed conditional', which sounds delightfully obscure, but simply means that the *probability of A given B* is confused with the *probability of B given A*.

or a clock. But how do we measure probability? There is no probability-ometer. It is as if probability is some 'virtual' quantity, which we can put a number on, but can never directly measure.

Even more worrying is to ask the rather obvious question: what does probability mean anyway? What's a good definition? This may seem pedantic, but the philosophy of probability is both a gripping topic in itself and also has a major role in practical applications of statistics.

Don't expect a neat consensus from the 'experts'. They may agree on the mathematics of probability, but philosophers and statisticians have come up with all sorts of different ideas for what these elusive numbers actually mean, and argue intensively over them. Some popular suggestions include:

- *Classical probability*: This is what we are taught in school, based on the symmetries of coins, dice, packs of cards, and so on, and can be defined as, 'The ratio of the number of outcomes favouring the event divided by the total number of possible outcomes, assuming the outcomes are all equally likely.' For example, the probability of throwing a 'one' on a balanced die is 1/6, since there are six faces. But this definition is somewhat circular, as we need to have a definition of 'equally likely'.
- *'Enumerative' probability*:* Suppose there are three white socks and four black socks in a drawer, and we take a sock at random, what is the probability of drawing a white sock? It is 3/7, obtained by enumerating the

---

* I am grateful to Philip Dawid for, apparently, inventing this term.

opportunities. Many of us have had to suffer questions like this in school, and it is essentially an extension of the classical idea discussed above, requiring the idea of a 'random choice' from a physical set of objects. We have been using this idea extensively already, when describing a data-point as being picked at random from a population.

- *'Long-run frequency' probability*: This is based on the proportion of times an event occurs in an infinite sequence of identical experiments, exactly as we found when we simulated the Chevalier's games. This may be reasonable (at least theoretically) for infinitely repeatable events, but what about unique occasions such as horse-racing, or tomorrow's weather? In fact almost any realistic situation is not, even in principle, infinitely repeatable.

- *Propensity or 'chance'*: This is the idea that there is some objective tendency of the situation to produce an event. This is superficially attractive – if you were an all-knowing being, maybe you could say there was a partic-ular probability of your bus arriving soon, or of being hit by a car today. But it seems to provide no basis for we mortals to estimate this rather metaphysical 'true chance'.

- *Subjective or 'personal' probability*: This is a specific person's judgement about a specific occasion, based on their current knowledge, and is roughly interpreted in terms of the betting odds (for small stakes) that they would find reasonable. So if I will be given £1 if I can juggle three balls for five minutes, and I am willing to offer a (non-repayable) 60p stake on the bet, then my probability for that event is 0.6.

Different 'experts' have their own preference among these alternatives, but personally I prefer the final interpretation – subjective probability. This means I take the view that any numerical probability is essentially *constructed* according to what is known in the current situation – indeed probability doesn't really 'exist' at all (except possibly at the subatomic level). This approach forms the basis for the **Bayesian** school of statistical inference, which we will explore in detail in Chapter 11.

But fortunately you don't have to agree with my (rather controversial) position that numerical probabilities do not objectively exist. It is fine to assume that coins and other randomizing devices are objectively random, in the sense that they give rise to data that are so unpredictable as to be indistinguishable from those we would expect to arise from 'objective' probabilities. So we generally act *as if* the observations are random, even when we know that this is not strictly true. The most extreme examples of this are pseudo-random-number generators, which are in fact based on logical and completely predictable calculations. They contain no randomness whatsoever, but their mechanism is so complex that they are in practice indistinguishable from truly random sequences – say, those obtained from a source of subatomic particles.*

This somewhat bizarre ability to act as if something is true, when you know it really isn't, would usually be

---

* This assumes that the pseudo-random-number generator is well designed, and that the intended use of these numbers is for statistical modelling or similar. They are not good enough for cryptographic applications, where the predictability could be used to break an encryption.

considered dangerously irrational. However, it will come in handy when it comes to using probability as a basis for the statistical analysis of data.

We now come to the crucial but difficult stage of laying out the general connection between probability theory, data and learning about whatever target population we are interested in.

Probability theory naturally comes into play in what we shall call situation 1:

1. When the data-point can be considered to be *generated* by some randomizing device, for example when throwing dice, flipping coins, or randomly allocating an individual to a medical treatment using a pseudo-random-number generator, and then recording the outcomes of their treatment.

But in practice we may be faced with situation 2:

2. When a pre-existing data-point is *chosen* by a randomizing device, say when selecting people to take part in a survey.

And much of the time our data arises from situation 3:

3. When there is no randomness at all, but we act as if the data-point were in fact generated by some random process, for example in interpreting the birth weight of our friend's baby.

Most expositions do not make these distinctions clear: probability is generally taught using randomizing devices (situation 1) and statistics is taught through the idea of 'random sampling' (situation 2), but in fact the majority of applications of statistics do not involve any random devices or random sampling whatsoever (situation 3).

But first consider situations 1 and 2. Just before we operate the randomizing device, we assume we have a set of possible results that might be observed, together with their respective probabilities – for example a coin can be heads or tails, each with probability of ½. If we associate each of these possible outcomes with a quantity, say in this case 0 for tails and 1 for heads, then we say we have a **random variable** with a probability distribution. In situation 1, the randomizing device ensures the observation is generated at random from this distribution, and when it is observed, the randomness has gone and all these potential futures have collapsed down on to the actual observation.* Similarly, in situation 2, if we draw an individual at random and, say, measure their income, then we have essentially drawn an observation at random from a population distribution of incomes.

So probability is clearly relevant when we have a randomizing device. But most of the time we simply consider all the measurements available to us at the time, which may have been collected informally, or, as we saw in Chapter 3, even represent every possible observation: think of survival rates for children's heart surgery at different hospitals or all examination results for British children – both of

---

* This can be considered analogous to the situation in quantum mechanics, in which the current state of, say, an electron is defined as a wave function that collapses down to a single state when it is actually observed.

these comprise all the data available, and there has been no random sampling.

In Chapter 3 we discussed the idea of a *metaphorical* population, comprising the possible eventualities that might have occurred, but mainly didn't. We now need to brace ourselves for an apparently irrational step: we need to act *as if* data were generated by a random mechanism from this population, even though we know full well that it was not.

If We Observe Everything, Where Does Probability Come In?

> How often do we expect to see seven or more separate homicide incidents in England and Wales in a single day?

When extreme events happen in close succession, such as multiple plane crashes or natural disasters, there is a natural propensity to feel they are in some sense linked. It then becomes important to work out just how unusual such events are, and the following example shows how we can make such a call.

To assess how rare a 'cluster' of at least seven homicides in a day might be, we can examine data for the three years (1,095 days) between April 2014 and March 2016, in which there were 1,545 homicide incidents in England and Wales, an average of $1,545/1,095 = 1.41$ per day.* Over this period there

---

* A 'homicide incident' is where the same person (or group of persons) is suspected of committing one or more related homicides. So a mass shooting or terrorist attack would count as one incident.

were no days with seven or more incidents, but it would be very naïve to therefore conclude that such an occurrence was impossible. If we can build a reasonable probability distribution for the number of homicides per day, then we can answer the question posed.

But what is the justification for building a probability distribution? The number of homicides recorded each day in a country is simply a fact – there has been no sampling, and there is no explicit random element generating each unfortunate event. Just an immensely complex and unpredictable world. But whatever our personal philosophy behind luck or fortune, it turns out that it is useful to act *as if* these events were produced by some random process driven by probability.

It might be useful to imagine that at the start of each day we have a large population of people, each of whom has a very small possibility of being a homicide victim. Data of this kind can be represented as observations from a **Poisson distribution**, which was originally developed by Siméon Denis Poisson in France in the 1830s to represent the pattern of wrongful convictions per year. Since then it has been used to model everything from the number of goals scored by a football team in a match or the number of winning lottery tickets each week, to the number of Prussian officers kicked to death by their horses each year. In each of these situations there is a very large number of opportunities for an event to happen, but each with a very low chance of occurrence, and this gives rise to the extraordinarily versatile Poisson distribution.

Whereas the normal (or Gaussian) distribution in Chapter 3 required two parameters – the population mean and standard deviation – the Poisson distribution depends only on its mean. In our current example this is the expected number of homicide incidents each day, which we assume to be 1.41, the average number each day over this three-year period. We should, though, carefully check whether the Poisson is a reasonable assumption, so that it is reasonable to act as if the number of homicides each day were a random observation drawn from a Poisson distribution with mean 1.41.

For example, just from knowing this average, we can use the formula for the Poisson distribution, or standard software, to calculate that there would be a probability of 0.01134 of exactly five homicides occurring in a day, which means that over 1,095 days we would expect $1,095 \times 0.01134 = 12.4$ days on which there were precisely five homicide incidents. Amazingly, the actual number of days over a three-year period on which there were five homicides was . . . 13.

Figure 8.5 compares the expected distribution of the daily number of homicide incidents based on a Poisson assumption, and the actual empirical data distribution over these 1,095 days – the match is very close indeed, and in Chapter 10 I will show how to test formally whether the Poisson assumption is justified.

In answer to the question posed at the start of this section, we can calculate from the Poisson distribution the probability of getting seven or more incidents in a day, which turns out to be 0.07%, and means that we can expect such an event to happen on average every 1,535 days, or roughly once every four years. We can conclude that this event is fairly unlikely to happen in the normal run of things, but is not impossible.

The fit of this mathematical probability distribution to the

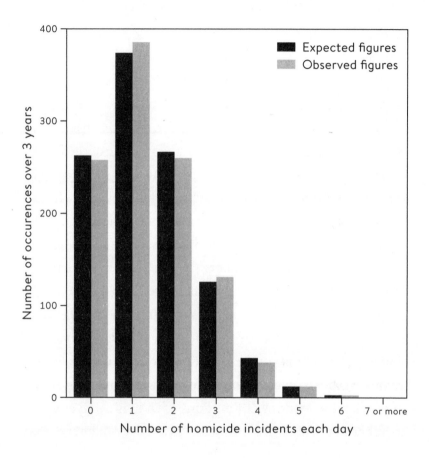

Figure 8.5
Observed and expected (assuming a Poisson distribution) daily number of recorded homicide incidents, 2014 to 2016, England and Wales.[3]

empirical data is almost disturbingly good. Even though there is a unique story behind every one of these tragic events, most of which are unpredictable, the data act as if they were actually generated by some known random mechanism. One possible view is to think that other people could have been murdered, but they weren't – we have observed one of many possible worlds that could have occurred, just as when we flip coins we observe one of the many possible sequences.

Adolphe Quetelet was an astronomer, statistician and sociologist in Belgium in the mid 1800s, and was one of the first to draw attention to the astonishing predictability of overall patterns made up of individually unpredictable events. He was intrigued by the occurrence of normal distributions in natural phenomena, such as the birth-weight distribution in Chapter 3, and coined the idea of '*l'homme moyen*' (the average man), who took on the mean value of all these characteristics. He developed the idea of 'social physics', since the regularity of societal statistics seemed to reflect an almost mechanistic underlying process. Just as the random molecules of a gas come together to make predictable physical properties, so the unpredictable workings of millions of individual lives come together to produce, for example, national suicide rates that barely change from year to year.

Fortunately we don't have to believe that events are actually driven by pure randomness (whatever that is). It is simply that an assumption of 'chance' encapsulates all the inevitable unpredictability in the world, or what is sometimes termed *natural variability*. We have therefore established that probability forms the appropriate mathematical foundation for both 'pure' randomness, which occurs with subatomic particles, coins, dice, and so on; and 'natural', unavoidable variability, such as in birth weights, survival after surgery, examination results, homicides, and every other phenomenon that is not totally predictable.

In the next chapter we come to a truly remarkable development in the history of human understanding: how these two aspects of probability can be brought together to provide a rigorous basis for formal statistical inference.

Summary

- The theory of probability provides a formal language and mathematics for dealing with chance phenomena.
- The implications of probability are not intuitive, but insights can be improved by using the idea of expected frequencies.
- The ideas of probability are useful even when there is no explicit use of a randomizing mechanism.
- Many social phenomena show a remarkable regularity in their overall pattern, while individual events are entirely unpredictable.

Putting Probability and Statistics Together

Warning. This is perhaps the most challenging chapter in this book, but persevering with this important topic will give you valuable insights into statistical inference.

> In a random sample of 100 people, we find that 20 are left-handed. What can we say about the proportion of the population who are left-handed?

In the last chapter we discussed the idea of a random variable – a single data-point drawn from a probability distribution described by parameters. But we are seldom interested in just one data-point – we generally have a mass of data which we summarize by determining means, medians and other statistics. The fundamental step we will take in this chapter is to consider those statistics as themselves being random variables, drawn from their own distributions.

This is a big advance, and one that has not only challenged generations of students of statistics, but also generations of statisticians who have tried to work out what distributions we should assume these statistics are drawn from. And given the discussion of the bootstrap in Chapter 7, it would be reasonable to ask why we need all that mathematics, when we can work out uncertainty intervals and so on using simulation-based bootstrap approaches. For example, the question posed at the start of this chapter could be answered by taking our observed data of 20 left-handed and 80 right-handed individuals, and repeatedly resampling 100 observations from this data set, with replacement, and looking at the distribution of the observed proportion of left-handed people.

But these simulations are clumsy and time-consuming, especially with large data sets, and in more complex circumstances it is not straightforward to work out what should be simulated. In contrast, formulae derived from probability theory provide both insight and convenience, and always lead to the same answer since they don't depend on a particular simulation. But the flip side is that this theory relies on assumptions, and we should be careful not to be deluded by the impressive algebra into accepting unjustified conclusions. We will explore this in more detail later, but first, having already appreciated the value of the normal and Poisson, we need to introduce another important probability distribution.

Suppose we draw samples of different sizes from a population containing exactly 20% left- and 80% right-handed people, and calculate the probability of observing different possible proportions of left-handers. Of course this is the wrong way round – we want to use the known sample to learn about the unknown population – but we can only get to this conclusion by first exploring how a known population gives rise to different samples.

The simplest case is a sample of one, when the observed proportion must be either 0 or 1 depending on whether we select a right- or left-hander – and these events occur with probability of 0.8 and 0.2 respectively. The resulting probability distribution is shown in Figure 9.1(a).

If we take two individuals at random, then the proportions of left-handers will either be 0 (both right-handers), 0.5 (one of each) or 1 (both left-handers). These events will occur with probabilities 0.64, 0.32 and 0.04 respectively,* and this probability distribution is shown in Figure 9.1(b). Similarly we can use probability theory to work out the probability distribution for the observed proportions of left-handers in the 5-, 10-, 100- and 1,000-person samples, which are all shown in Figure 9.1. These distributions are based on what is known as the **binomial distribution**, and can also tell us the probability, for example, of getting at least 30% left-handed people if we sample 100, known as a tail-area.

The mean of a random variable is also known as its **expectation**, and in all these samples we expect a proportion of 0.2 or 20%: all the distributions shown in Figure 9.1 have 0.2 as their mean. The standard deviation for each is given by a formula which depends on the underlying proportion, in this case 0.2, and the sample size. Note that the standard deviation of a statistic is generally termed the **standard error**, to distinguish it from the standard deviation of the population distribution from which it derives.

---

* To derive this distribution, we could calculate the probability of two left-handers as 0.2 × 0.2 = 0.04, the probability of two right-handers as 0.8 × 0.8 = 0.64, and so the probability of one of each must be 1 − 0.04 − 0.64 = 0.32.

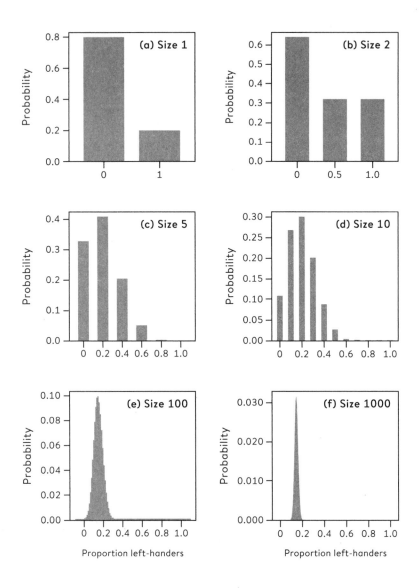

Figure 9.1
The probability distribution of the observed proportion of left-handers in random samples of 1, 2, 5, 10, 100 and 1,000 people, where the true underlying proportion of left-handers in the population is 0.2. The probability of getting at least 30% left-handers in the sample is found by adding all the probability in the bars to the right of 0.3.

Figure 9.1 has some distinctive features. First, the probability distributions tend to a regular, symmetric, normal shape as the sample size increases, just as we observed using bootstrap simulations. Second, the distributions get tighter as the sample size increases. The following example shows how a simple application of these ideas can be used to rapidly identify whether a statistical claim is reasonable or not.

Do some areas of the UK really have three times the bowel-cancer death rates than others?

The headline on the respected BBC news website in September 2011 was alarming: 'Threefold Variation in UK Bowel Cancer Death Rates'. The article went on to explain that different areas in the UK had starkly different death rates from bowel cancer, with a commentator suggesting it was 'extremely important for local NHS organizations to examine information for their own areas and use it to inform potential changes in delivery of services'.

A threefold difference sounds extraordinarily dramatic. But when the blogger Paul Barden came across the article, he wondered, 'Do people in different parts of the country really face such large and important differences in their risk of dying from bowel cancer? What would cause such a discrepancy?' He found it so implausible he decided to investigate. Admirably, the data was openly available online and he found that it did substantiate what the BBC piece had claimed: in 2008 there was more than a threefold variation between the annual death rates of people with bowel cancer. It ranged from 9 per 100,000 people in Rossendale in Lancashire to 31 per 100,000 inhabitants of Glasgow City.[1]

But this was not the end of his investigation. He then plotted the death rates against the population in each district, which gave the picture shown in Figure 9.2. It is clear that the points (all apart from the extreme example of Glasgow City) form a sort of funnel shape, in which the differences between districts get larger as their population gets smaller. Paul then added **control limits** which show where we would expect the points to land if the differences between the observed rates were just due to natural and unavoidable variability in the numbers that die of bowel cancer each year, rather than due to any systematic variation in the underlying risks experienced in different districts. These control limits are obtained from assuming that the number of bowel cancer deaths in each area are an observation from a binomial distribution with a sample size equal to the adult population of the area, and an underlying probability of 0.000176 that any particular person would die from bowel cancer each year: this is the average individual risk over the whole country. The control limits are set to contain 95% and 99.8% of the probability distribution respectively. This type of graph is called a **funnel plot** and is extensively used when examining multiple health authorities or institutions, since it permits the identification of outliers without creating spurious league tables.

The data fall within the control limits rather well, which means that differences between districts are essentially what we would expect by chance variability alone. Smaller districts have fewer cases and so are more vulnerable to the role

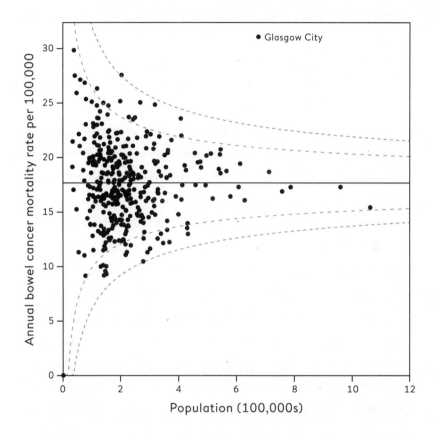

Figure 9.2

Annual bowel-cancer death rates per 100,000 population in 380 districts in the UK, plotted against the population of the district. The two sets of dashed lines indicate the regions in which we would expect 95% and 99.8% of districts to lie, if there were no real differences between the risks, and they are derived from an assumed underlying binomial distribution. Only Glasgow City shows any evidence of an underlying risk that is different from the average. This way of looking at the data is called a 'funnel plot'.

of chance, and therefore tend to have more extreme results – the rate in Rossendale was based on only 7 deaths, and so its rate could be drastically altered by just a few extra cases. So despite the BBC's dramatic headline, there is no big news story here – we would expect a threefold variability in the observed rates, even if the underlying risk in the different districts were precisely the same.

There is a crucial lesson in this simple example. Even in an era of open data, data science and data journalism, we still need basic statistical principles in order not to be misled by apparent patterns in the numbers.

This chart reveals that the only observation of any particular note is Glasgow City's outlying data-point. Is bowel cancer a particularly Scottish phenomenon? Is this data-point actually correct? More recent data for the period 2009–2011 reveals that bowel-cancer mortality for Greater Glasgow was 20.5 per 100,000 people, in Scotland overall it was 19.6, and in England it was 16.4: these findings both cast doubt on the specific Glasgow City value and show that Scotland has higher rates than England. Typically, conclusions from one problem-solving cycle raise more questions, and so the cycle starts over again.

The Central Limit Theorem

Individual data-points might be drawn from a wide variety of population distributions, some of which might be highly skewed, with long tails such as those of income or sexual partners. But we have now made the crucial shift to considering distributions of statistics rather than individual data-points, and these statistics will commonly be averages of some sort. We have already seen in Chapter 7 that the distribution of the sample means of bootstrap resamples tends to a well-behaved symmetric shape, whatever the shape of the original distribution of the data, and we can now go beyond this to a deeper and rather remarkable idea, established around 300 years ago.

The example of left-handers shows that the variability in the observed proportion gets smaller as the sample size increases – this is why the funnel in Figure 9.2 gets narrower around the mean. This is the classic **Law of Large Numbers**, which was established by Swiss mathematician Jacob Bernoulli in the early eighteenth century – a single coin flip, taking on the value 1 if a head occurs, and 0 if a tail, is said to be a Bernoulli trial and have a **Bernoulli distribution**. If you keep on flipping a balanced coin, carrying out more and more Bernoulli trials, then the proportion of each outcome will get closer and closer to 50% heads and 50% tails – we say the observed proportion converges to the true underlying chance of a head. Of course, early on in the sequence the ratio may be some way from 50:50, say after a run of heads, and the temptation is to believe that tails is somehow now 'due' so that the proportion gets balanced out – this is known as the 'gambler's fallacy' and is a psychological bias that (from personal experience) is rather difficult to overcome. But the coin has no memory – the key insight is that the coin cannot *compensate* for past imbalances, but simply *overwhelms* them by more and more new, independent flips.

In Chapter 3 we introduced the classic 'bell-shaped curve', also known as the normal or Gaussian distribution, where we showed it described well the distribution of birth weights in the US population, and argued that this was because birth weight depends on a huge number of factors, all of which have a little influence – when we add up all those small effects we get a normal distribution.

This is the reasoning behind what is known as the Central Limit Theorem, first proved in 1733 by French mathematician Abraham de Moivre for the particular case of the binomial distribution. But it is not just the binomial distribution that tends to a normal curve with increasing sample size – it is a remarkable fact that virtually *whatever* the shape of the population distribution from which each of the original measurements are sampled, for large sample sizes their average can be considered to be drawn from a normal curve.* This will have a mean that is equal to the mean of the original distribution and a standard deviation that has a simple relationship to the standard deviation of the original population distribution and, as already mentioned, is often known as the standard error.†

Apart from his work on wisdom of crowds, correlation, regression, and almost everything else, Francis Galton also considered it a true marvel that the normal distribution, then known as the Law of Frequency of Error, should arise in an orderly way out of apparent chaos:

> I know of scarcely anything so apt to impress the imagination as the wonderful form of cosmic order

---

* There are important exceptions to this – some distributions have such long, 'heavy' tails that their expectations and standard deviations do not exist, and so averages have nothing to converge to.

† If we can assume that all our observations are independent and come from the same population distribution, the standard error of their average is just the standard deviation of the population distribution divided by the square root of the sample size.

expressed by the 'Law of Frequency of Error'. The law would have been personified by the Greeks and deified, if they had known of it. It reigns with serenity and in complete self-effacement, amidst the wildest confusion. The huger the mob, and the greater the apparent anarchy, the more perfect is its sway. It is the supreme law of Unreason. Whenever a large sample of chaotic elements are taken in hand and marshalled in the order of their magnitude, an unsuspected and most beautiful form of regularity proves to have been latent all along.

He was right – it really is an extraordinary law of nature.

How Does This Theory Help Us Work Out the Accuracy of Our Estimates?

All this theory is fine for proving things about distributions of statistics based on data drawn from known populations, but that is not what we are mostly interested in. We have to find a way of reversing the process: instead of going from known populations to saying something about possible samples, we need to go from a single sample back to saying something about a possible population. This is the process of inductive inference outlined in Chapter 3.

Suppose I have a coin, and I ask you for your probability that it will come up heads. You happily answer '50:50', or similar. Then I flip it, cover up the result before either of us sees it, and again ask for your probability that it is heads. If you are typical of my experience, you may, after a pause, rather grudgingly say '50:50'. Then I take a quick look at the coin, without showing you, and repeat the question. Again, if you are like most people, you eventually mumble '50:50'.

This simple exercise reveals a major distinction between two types of uncertainty: what is known as **aleatory uncertainty** *before* I flip the coin – the 'chance' of an unpredictable event – and **epistemic uncertainty** *after* I flip the coin – an expression of our personal ignorance about an event that is fixed but unknown. The same difference exists between a lottery ticket (where the outcome depends on chance) and a scratch card (where the outcome is already decided, but you don't know what it is).

Statistics are used when we have epistemic uncertainty about some quantity of the world. For example, we conduct a survey when we don't know the true proportion in a population that consider themselves religious, or we run a pharmaceutical trial when we don't know the true average effect of a drug. As we have seen, these fixed but unknown quantities are called parameters and are often given a Greek letter.* Just like my coin-flipping example, *before* we do these experiments we have aleatory uncertainty about what the outcomes may be, because of the random sampling of individuals or the random allocation of patients to the drug or a dummy tablet. Then *after* we have done the study and got the data, we use this probability model to get a handle on our current epistemic uncertainty, just as you were eventually prepared to say '50:50' about the covered-up coin. So probability theory, which tells us what to expect in the future, is used to tell us what we can learn from what we have observed in the

---

* We shall see in Chapter 12 that practitioners of Bayesian statistics are happy using probabilities for epistemic uncertainty about parameters.

past. This is the (rather remarkable) basis for statistical inference.

The procedure for deriving an uncertainty interval around our estimate, or equivalently a margin of error, is based on this fundamental idea. There are three stages:

1. We use probability theory to tell us, for any particular population parameter, an interval in which we expect the observed statistic to lie with 95% probability. These are 95% prediction intervals, such as those displayed in the inner funnel in Figure 9.2.
2. Then we observe a particular statistic.
3. Finally (and this is the difficult bit) we work out the range of possible population parameters for which our statistic lies in their 95% prediction intervals. This we call a '95% **confidence interval**'.
4. This resulting confidence interval is given the label '95%' since, with repeated application, 95% of such intervals should contain the true value.[*]

All clear? If it isn't, then please be reassured that you have joined generations of baffled students. Specific formulae are provided in the Glossary, but the details are less important than the fundamental principle: a confidence interval is the range of population parameters for which our observed statistic is a plausible consequence.

---

[*] Strictly speaking, a 95% confidence interval does *not* mean there is a 95% probability that this particular interval contains the true value, although in practice people often give this incorrect interpretation.

Calculating Confidence Intervals

The principle of confidence intervals was formalized in the 1930s at University College London by Jerzy Neyman, a brilliant Polish mathematician and statistician, and Egon Pearson, Karl Pearson's son.* The work of deriving the necessary probability distributions of estimated correlation coefficients and regression coefficients had been going on for decades beforehand, and in standard academic statistics courses the mathematical details of these distributions would be provided, and even derived from first principles. Fortunately the results of all these labours are now encapsulated in statistical software, and so practitioners can focus on the essential issues and not be distracted by complex formulae.

We saw in Chapter 7 how bootstrapping could be used to get 95% intervals for the gradient of Galton's regression of daughters' on mothers' heights. It is far easier to obtain exact intervals that are based on probability theory and provided in standard software, and Table 9.1 shows they give very similar results. The 'exact' intervals based on probability theory require more assumptions than the bootstrap approach, and strictly speaking would only be precisely correct if the underlying population distribution were normal. But the Central Limit Theorem means that with such a large sample size it is reasonable to assume our estimates have got normal distributions and so the exact intervals are acceptable.

---

* Both of whom I had the pleasure of knowing in their more advanced years.

	Gradient of regression of offspring on parent		
	Estimate	Standard error	95% interval
Exact	0.33	0.05	0.23 to 0.42
Bootstrap	0.33	0.06	0.22 to 0.44

Table 9.1
Estimates of the regression coefficient summarizing the relationship between mothers' and daughters' heights, with exact and bootstrap standard errors and 95% confidence intervals – the bootstrap is based on 1,000 resamples.

It is conventional to use 95% intervals, which are generally set as plus or minus two standard errors, but narrower (for example, 80%) or wider (for example, 99%) intervals are sometimes adopted.* The US Bureau of Labor Statistics use 90% intervals for unemployment, whereas the UK Office for National Statistics use 95%: it is essential to be clear which is being used.

Margins of Error from Surveys

When it is clear that a claim is based on a survey, such as an opinion poll, it is standard practice to report a margin of error. The unemployment statistics introduced in Chapter 7 had surprisingly large margins of error, with the estimated change of 3,000 having a margin of error of $\pm$ 77,000. This has a material effect on the interpretation of the original number – in this case the margin of error reveals that we cannot even be sure if unemployment has gone up or down.

A simple rule of thumb is that, if you are estimating the percentage of people who prefer, say, coffee to tea for breakfast, and you ask a random sample from a population, then your margin of error (in %) is at most plus or minus 100 divided by the square root of the sample size.[2] So for a survey of 1,000 people (the industry standard), the margin of error is generally quoted as $\pm$ 3%:† if 400 of them said they preferred

---

* More precisely, 95% confidence intervals are often set as plus or minus 1.96 standard errors, based on assuming a precise normal sampling distribution for the statistic.

† With 1,000 participants, the margin of error (in %) is at most $\pm 100/\sqrt{1,000} = 3\%$. Surveys may have more complex designs than taking a simple random sample from a population, but the margins of error are not strongly affected.

coffee, and 600 of them said they preferred tea, then you could roughly estimate the underlying percentage of people in the population who prefer coffee as 40 ± 3%, or between 37% and 43%.

Of course, this is only accurate if the polling company really did take a random sample, and everyone replied, and they all had an opinion either way and they all told the truth. So although we can calculate margins of error, we must remember that they only hold if our assumptions are roughly correct. But can we rely on these assumptions?

Should We Believe Margins of Error?

Before the UK general election in June 2017, numerous opinion polls were published on the voting intentions of around 1,000 respondents. If these had been perfect random polls to which the participants had given truthful responses, then the margin of error of each should have been at most ± 3%, and so the variability of the polls around their running average should have been in that range, since they were all supposedly measuring the same underlying population. But Figure 9.3, based on a graphic used by the BBC, shows that the variability was far larger than this, meaning that the margins of error could not be correct.

We have already seen many of the reasons why surveys can be inaccurate, on top of the inevitable (and quantifiable) margin of error due to random variability. In this case the excess variability might be blamed on the sampling methods, in particular the use of telephone polls with a very low response rate, perhaps between 10% and 20%, and mainly using landlines. My personal, rather sceptical heuristic is that

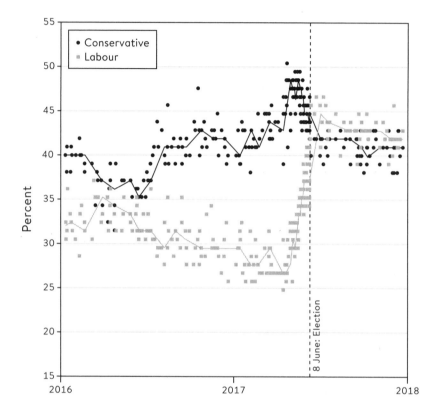

Figure 9.3

The style of visualization of opinion poll data used by the BBC before the UK general election on 8 June 2017.[3] The trend line is the median of the previous seven polls. Each poll was usually based on 1,000 respondents, and therefore claimed to have a margin of error of at most ±3%. But the variability between polls far exceeded this margin. Parties other than Labour and Conservative are not shown.

any quoted margin of error in a poll should be doubled to allow for systematic errors made in the polling.

We might not expect complete accuracy in pre-election polls, but we would expect more from scientists trying to measure physical facts about the world such as the speed of light. But there is a long history of claimed margins of error from such experiments later being found to be hopelessly inadequate: in the first part of the twentieth century, the uncertainty intervals around the estimates of the speed of light did not include the current accepted value.

This has led organizations that work on *metrology*, the science of measurement, to specify that margins of error should always be based on two components:

- Type A: the standard statistical measures discussed in this chapter, which would be expected to reduce with more observations.
- Type B: systematic errors that would not be expected to reduce with more observations, and have to be handled using non-statistical means such as expert judgement or external evidence.

These insights should encourage us to have humility about the statistical methods we can bring to a single data source. If there are fundamental problems with the way the data have been collected then no amount of clever methods can eliminate these biases, and we have to use our background knowledge and experience to temper our conclusions.

What Happens When We Have All the Data There Is?

It seems natural to use probability theory to put margins of error around survey results, since individuals have been randomly sampled from a larger population, so there is a clear way in which chance comes into the production of the data. But again we ask the question: what if the statistic being quoted is a complete count of everything that has happened? For example, each year a country will count its homicides. Assuming there is no error in the actual counting (and that it is agreed what a 'homicide' means), then these are simply descriptive statistics with no margin of error.

But suppose we wanted to make a claim about underlying trends over time, say 'the murder rate in the UK is going up'. For example, the UK Office for National Statistics reported that there were 497 homicides between April 2014 and March 2015, and 557 in the following year. Certainly the number of homicides has gone up, but we know that the number of murders varies from year to year for no apparent reason, so does this represent a real change in the underlying annual homicide rate? We want to make an inference about this unknown quantity, and so we need a probability model for our observed homicide counts.

Fortunately, we saw in the last chapter that the number of homicides each day act as if they are random observations drawn with a Poisson distribution from a metaphorical population of alternative possible histories. This in turn means the total over the whole year can be considered as a single observation from a Poisson distribution with mean m equal to the (rather hypothetical) 'true' underlying annual rate. Our interest is whether m changes from year to year.

The standard deviation of this Poisson distribution is the square root of m, written $\sqrt{m}$, which is also the standard error of our estimate. This would allow us to create a confidence interval, if only we knew m. But we don't (that's the whole point of the exercise). Consider the 2014–2015 period, when there were 497 homicides, which is our estimate for the underlying rate m that year. We can use this estimate for m to estimate the standard error $\sqrt{m}$ as $\sqrt{497} = 22.3$. This gives a margin of error of $\pm 1.96 \times 22.3 = \pm 43.7$. So we can finally get to our approximate 95% interval for m as $497 \pm 43.7 = 453.3$ to 540.7. Since 95% confidence intervals are often assumed to be plus or minus 1.96 standard errors, this means that we can be 95% confident that during this period the true underlying homicide rate lies between 453 and 541 per year.

Figure 9.4 shows the observed number of homicides in England and Wales between 1998 and 2016, with 95% confidence intervals for the underlying rate. It is clear that although there is inevitable variation between the annual counts, the confidence intervals show that we need to be cautious in drawing conclusions about changes over time. For example, the 95% interval around the 2015–2016 count of 557 goes from 511 to 603, with a substantial overlap with the confidence interval for the previous year.

So how can we decide if there has been a real change in the underlying risk of being a homicide victim, or whether the observed changes can just be put down to unavoidable chance variation? If the confidence intervals do not overlap, then we can certainly be at least 95% confident there has been a real change. But this is a rather stringent criterion,

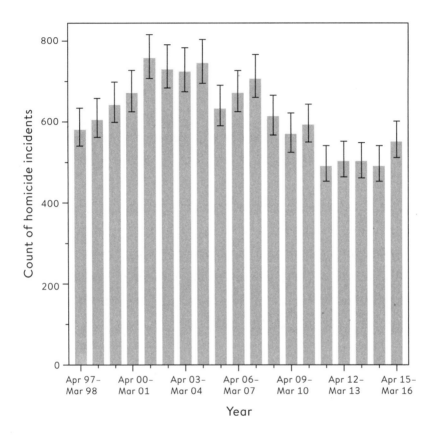

Figure 9.4
Number of homicides each year in England and Wales between 1998 and 2016, and 95% confidence intervals for the underlying 'true' homicide rate.[4]

and we should really create a 95% interval for the change in underlying rates. Then if this interval includes 0, we cannot be confident there has been a real change.

There was an increase of $557 - 497 = 60$ in the number of homicides between 2014–2015 and 2015–2016. It turns out that a 95% confidence interval around this observed change runs from −4 to +124, which (just) includes 0. Technically this means that we cannot conclude with 95% confidence that the underlying rate has changed, but since we are right on the margin it would be unreasonable to proclaim that there has been no change at all.

The confidence intervals around the homicide counts in Figure 9.4 are of a totally different nature to margins of error around, say, unemployment figures. The latter are an expression of our epistemic uncertainty about the actual number of people unemployed, while the intervals around homicide counts are not expressing uncertainty about the actual number of homicides – we assume these have been correctly counted – but the underlying risks in society. These two types of interval may look similar, and even use similar mathematics, but they have fundamentally different interpretations.

There has been some challenging material in this chapter, which is unsurprising as it has essentially laid out the whole formal foundation for statistical inference based on probability modelling. But the effort is worthwhile, as we can now use this structure to go beyond basic description and estimation of characteristics of the world, and start seeing how statistical modelling can help answer important questions about how the world actually works, and so provide a firm basis for scientific discoveries.

Summary

- Probability theory can be used to derive the sampling distribution of summary statistics, from which formulae for confidence intervals can be derived.
- A 95% confidence interval is the result of a procedure that, in 95% of cases in which its assumptions are correct, will contain the true parameter value. It cannot be claimed that a specific interval has 95% probability of containing the true value.
- The Central Limit Theorem implies that sample means and other summary statistics can be assumed to have a normal distribution for large samples.
- Margins of error usually do not incorporate systematic error due to non-random causes – external knowledge and judgement is required to assess these.
- Confidence intervals can be calculated even when we observe all the data, which then represent uncertainty about the parameters of an underlying metaphorical population.

CHAPTER 10

Answering Questions and Claiming Discoveries

Are more boys born than girls?

John Arbuthnot, a doctor who became physician to Queen Anne in 1705, set out to determine the answer to this question. He examined data on London baptisms for the 82 years between 1629 and 1710, and his results are shown in Figure 10.1 in terms of what is now known as the sex ratio, which is the number of boys born per 100 girls.

He found there had been more males than females baptized in every year, with an overall sex ratio of 107, varying between 101 and 116 over the period. But Arbuthnot wanted to claim a more general law, and so argued that if there were really no difference in the underlying rates of boys and girls being born, then each year there would be a 50:50 chance that more boys than girls were born, or more girls than boys, just like flipping a coin.

But to get an excess of boys in every year would then be like flipping a fair coin 82 times in a row, and getting heads every time. The probability of this happening is $1/2^{82}$, which is a very small number indeed, with 24 zeros after the decimal

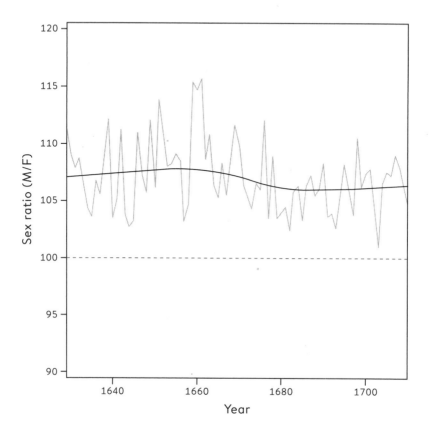

Figure 10.1
The sex ratio (number of boys per 100 girls) for London baptisms between 1629 and 1710, published by John Arbuthnot in 1710. The solid line represents an equal number of boys and girls; the curve is fitted to the empirical data. In all years there were more baptized boys than girls.

place. If we observed this in a real experiment, we would confidently claim that the coin was not fair. Similarly Arbuthnot concluded that some force was at work that produced more boys, which he thought must be to counter the greater mortality of males: 'To repair that Loss, provident Nature, by the Disposal of its wise Creator, brings forth more Males than Females; and that in almost a constant proportion.'[1]

Arbuthnot's data have been subject to repeated analysis, and although there may be counting errors, and only Anglican baptisms were included, his basic finding still holds: the 'natural' sex ratio is now considered to be around 105, meaning 21 boys are born for every 20 girls. The title of the article he published uses his data as direct statistical evidence for the existence of supernatural intervention: 'An Argument for Divine Providence, Taken from the Constant Regularity Observed in the Births of Both Sexes'. Whether or not this is a justified conclusion, and although he was unaware of it at the time, he had entered history by conducting the world's first test of statistical significance.

We have reached perhaps the most important part of the problem-solving cycle, in which we seek answers to specific questions about how the world works. For example:

1. Do the daily number of homicides in the UK follow a Poisson distribution?
2. Has the unemployment rate in the UK changed in the last quarter?
3. Does taking statins reduce the risk of heart attacks and strokes in people like me?

4. Are mothers' heights associated with their sons' heights, once the fathers' heights are taken into account?

5. Does the Higgs boson exist?

This list shows that very different kinds of question can be asked, ranging from the transient to the eternal:

1. Homicides and the Poisson distribution: a general rule that is not of great interest to the public, but helps to answer whether there has been a change in underlying rates.

2. Changing unemployment: a specific question concerning a particular time and place.

3. Statins: a scientific statement, but specific to a group.

4. Mothers' heights: possibly of general scientific interest.

5. Higgs boson: could change the basic ideas of the physical laws of the universe.

We have data that can help us answer some of these questions, with which we have already done some exploratory plotting and drawn some informal conclusions about an appropriate statistical model. But we now come to a formal aspect of the Analysis part of the PPDAC cycle, generally known as **hypothesis testing**.

What Is a 'Hypothesis'?

A hypothesis can be defined as a proposed explanation for a phenomenon. It is not the absolute truth, but a provisional, working assumption, perhaps best thought of as a potential suspect in a criminal case.

When discussing regression in Chapter 5, we saw the claim that

$$\text{observation} = \text{deterministic model} + \text{residual error}.$$

This represents the idea that statistical models are mathematical representations of what we observe, which combine a deterministic component with a 'stochastic' component, the latter representing unpredictability or random 'error', generally expressed in terms of a probability distribution. Within statistical science, a hypothesis is considered to be a particular assumption about one of these components of a statistical model, with the connotation of being provisional, rather than 'the truth'.

Why Do We Need Formal Testing of Null Hypotheses?

It is not just scientists who value discoveries – the delight in finding something new is universal. In fact it is so desirable that there is an innate tendency to feel we have found something when we have not. We previously used the term *apophenia* to describe the capacity to see patterns where they do not exist, and it has been suggested that this tendency might even confer an evolutionary advantage – those ancestors who ran away from rustling in the bushes without waiting to find out whether it was definitely a tiger may have been more likely to survive.

But while this attitude may be fine for hunter-gatherers, it cannot work in science – indeed, the whole scientific process is undermined if claims are just figments of our imagination.

There must be a way of protecting us against false discoveries, and hypothesis testing attempts to fill that role.

The idea of a **null hypothesis** now becomes central: it is the simplified form of statistical model that we are going to work with until we have sufficient evidence against it. In the questions listed above, the null hypotheses might be:

1. The daily number of homicides in the UK *do* follow a Poisson distribution.
2. The UK unemployment rate *has remained unchanged* over the last quarter.
3. Statins *do not* reduce the risk of heart attacks and strokes in people like me.
4. Mothers' heights *have no effect* on sons' heights, once fathers' heights are taken into account.
5. The Higgs boson *does not* exist.

The null hypothesis is what we are willing to assume is the case until proven otherwise. It is relentlessly negative, denying all progress and change. But this does not mean that we actually believe the null hypothesis is literally true: it should be clear that none of the hypotheses listed above could plausibly be precisely correct (except possibly the non-existence of the Higgs boson). So we can never claim that the null hypothesis has been actually proved: in the words of another great British statistician, Ronald Fisher, 'the null hypothesis is never proved or established, but is possibly disproved, in the course of experimentation. Every experiment may be said to exist only in order to give the facts a chance of disproving the null hypothesis.'[2]

There is a strong analogy with criminal trials in the English legal system: a defendant can be found guilty, but nobody is ever found innocent, simply not proven to be guilty. Similarly we shall find that we may reject the null hypothesis, but if we don't have sufficient evidence to do so, it does not mean that we can accept it as truth. It is just a working assumption until something better comes along.

> Fold your arms. Is your left or right arm on top? Studies show that around half the population put their right arm on top, around half their left. But is this associated with whether you are male or female?

Although this is perhaps not the most urgent scientific question, it is one that I investigated while teaching at the African Institute of Mathematical Sciences in 2013 – it was a good classroom exercise, and I was genuinely interested in the answer.* I obtained data from 54 post-graduate students who originated from all over Africa. Table 10.1 shows the total responses by gender and whether left or right arm was on top. This type of table is known as a cross-tabulation or a contingency table.

Overall, the majority placed their right arm on top (32/54 = 59%). However a higher proportion of females (9/14 = 64%) than males (23/40 = 57%) were 'right-armers': the observed difference in proportions is 64%−57% = 7%. In this case, the null hypothesis would be that there is truly no association

---

* Perhaps a more natural question is the relationship between arm-crossing and handedness, but there were too few left-handers to investigate this.

	Female	Male	Total
Left arm on top	5	17	22
Right arm on top	9	23	32
Total	14	40	54

Table 10.1
Cross-tabulation of the genders and arm-crossing behaviour for 54 post-graduate students.

whatsoever between arm-crossing and gender, in which case we would expect the observed difference in proportions between genders to be 0%. But of course inevitable random variability between people, even under this null assumption, means that the observed difference is not going to be precisely 0%. The crucial question is whether the observed difference of 7% is big enough to provide evidence against the null hypothesis.

To answer this, we need to know what kind of observed differences in proportions we would expect to happen simply due to random variation – that is, if the null hypothesis were actually true and arm-crossing was entirely independent of gender. More formally, is this observed difference of 7% compatible with the null hypothesis?*

This is a tricky yet crucial idea. When Arbuthnot was testing his null hypothesis that boys and girls were equally likely, he could easily work out that his observed data was not in the least compatible with the null assumption – the chance that boys would exceed girls in 82 years, if all that were operating were chance, was utterly tiny. In more complex situations it is not so straightforward to work out whether the data is compatible with the null hypothesis, but the following **permutation test** illustrates a powerful procedure that avoids complex mathematics.

Imagine all 54 students lined up in a row, with the first 14 being women and the next 40 being the men, and each is

---

* We could choose another statistic that summarizes associations, such as the odds ratio, but would get essentially the same result.

given a number from 1 to 54. Imagine that each of them also holds a ticket indicating whether they were left- or right-armers. Now imagine taking these 'arm-crossing' tickets, mixing them up in a hat, and dealing them out to the students at random. This is an example of how we would expect nature to work if the null hypothesis were true, since arm-crossing would then be completely unrelated to gender.

Even though arm-crossing behaviour has now been allocated at random, the proportion of right-armers will not be exactly the same for females and males, just by the play of chance, and we can calculate the observed difference in proportions for this random relabelling of the students. Then we could repeat this process of randomly allocating arm-crossing behaviour, say 1,000 times, and see what distribution of differences is generated. The results are shown in Figure 10.2(a), and show a scatter of observed differences – some favouring males, some females – centred on a difference of zero. The actual observed difference sits near the centre of this distribution.

An alternative approach, if we had rather a lot of time, would be to systematically work through all the possible permutations of the arm-crossing tickets, rather than just doing 1,000 simulations. Each of these would generate an observed difference in proportions of right-armedness between males and females, and plotting these would produce a smoother distribution than for just 1,000 simulations.

Unfortunately there is a vast number of such permutations, and even if calculated at a million a second, the number of years it would take to run through them all has 57 zeros after it.[3] Fortunately we don't have to perform these calculations

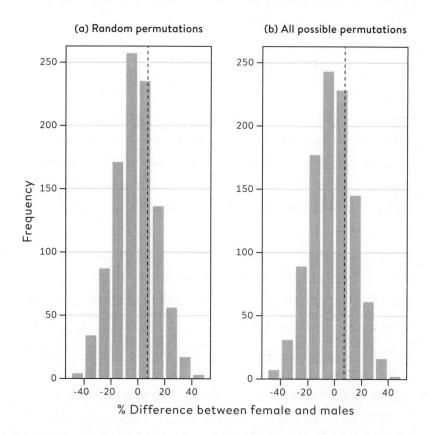

Figure 10.2
Empirical distribution of difference in proportions of women and men who crossed their arms with their right arm on top: (a) from 1,000 random permutations of arm-crossing, (b) from all possible equally likely permutations of the arm-crossing response. The observed difference in proportions (7%) is indicated by a vertical dashed line.

since the probability distribution for the observed difference in proportions under the null hypothesis can be worked out in theory, and is shown in Figure 10.2(b) – it is based on what is known as the **hypergeometric distribution,** which gives the probability for a particular cell in the table taking on each possible value under random permutations.

Figure 10.2 shows that the actual observed difference in proportions of right-armers (7% in favour of females) lies fairly near the centre of the distribution of observed differences that we would expect to see, if in truth there were no association at all. We need a measure to summarize how close to the centre our observed value lies, and one summary is the 'tail-area' to the right of the dashed line shown in Figure 10.2, which is 45% or 0.45.

This tail-area is known as a **P-value,** one of the most prominent concepts in statistics as practised today, and which therefore deserves a formal definition in the text: *A P-value is the probability of getting a result at least as extreme as we did, if the null hypothesis (and all other modelling assumptions) were really true.*

The issue, of course, is what do we mean by 'extreme'? Our current P-value of 0.45 is **one-tailed,** since it only measures how likely it is that we would have observed such an extreme value in favour of females, were the null hypothesis really true. This P-value corresponds to what is known as a **one-sided test.** But an observed proportion in favour of males would also have led us to suspect the null hypothesis did not hold. We should therefore also calculate the chance of getting an observed difference of at least 7%, in *either* direction. This is known as a **two-tailed** P-value, corresponding to a **two-sided test**. This total tail area turns out to be 0.89, and since this value is near one it indicates that the observed value is near the centre of the null distribution. Of course this could be seen immediately from Figure 10.2, but such graphs will not always be available and we need a number to formally summarize the extremeness of our data.

Arbuthnot provided the first recorded example of this process: under the null hypothesis that boys and girls were equally likely to be born, the probability of boys exceeding girls in all 82 years was $1/2^{82}$. This only defines extremeness in terms of boys exceeding girls, and we would also doubt the null hypothesis if girls exceeded boys, and so we should double this number to $1/2^{81}$ to give the probability of such an extreme result in either direction. So $1/2^{81}$ might be considered the first recorded two-sided P-value, although the term was not used for another two hundred and fifty years.

Incidentally, my small sample indicated no link between gender and arm-crossing, and other, more scientific studies have not found a relationship between arm-crossing behaviour and gender, handedness or any other feature.

Statistical Significance

The idea of **statistical significance** is straightforward: if a P-value is small enough, then we say the results are statistically significant. This term was popularized by Ronald Fisher in the 1920s and, in spite of the criticisms we shall see later, continues to play a major role in statistics.

Ronald Fisher was an extraordinary, but difficult, man. He was extraordinary because he is regarded as a pioneering figure in two distinct fields – genetics and statistics. Yet he had a notorious temper and could be extremely critical of anyone who he felt questioned his ideas, while his support for eugenics and his public criticism of the evidence for the link between smoking and lung cancer damaged his standing. His personal reputation has suffered as his financial connections with the tobacco industry have been revealed, but his scientific reputation is undiminished, as his ideas find repeated new applications in the analysis of large data sets.

As I mentioned in Chapter 4, Fisher developed the idea of randomization in agricultural trials while working at Rothamsted Experimental Station. He further illustrated the ideas of randomization in experimental design with his famous tea tasting test, in which a woman (thought to be a Muriel Bristol) claimed to be able to tell by tasting a cup of tea whether the milk had been added before or after tea was poured into a cup.

Four cups with milk first, and four with tea first, were prepared and the eight cups were presented in a random order; Muriel was told there were four of each, and had to guess which four had milk first. She is said to have got them all right, which another application of the hypergeometric distribution shows has a probability of 1 in 70 under the null hypothesis that she was guessing. This is an example of a P-value, and would by convention be considered small, and so the results could be declared to be statistically significant evidence that she could in fact tell whether the milk was put in first or not.

To summarize, I have described the following steps:

1. Set up a question in terms of a null hypothesis that we want to check. This is generally given the notation H_0.

2. Choose a test statistic that estimates something that, if it turned out to be extreme enough, would lead us to doubt the null hypothesis (often larger values of the statistic indicate incompatibility with the null hypothesis).

3. Generate the sampling distribution of this test statistic, were the null hypothesis true.

4. Check whether our observed statistic lies in the tails of this distribution and summarize this by the P-value: the probability, were the null hypothesis true, of observing such an extreme statistic. The P-value is therefore a particular tail-area.

5. 'Extreme' has to be defined carefully – if say both large positive and large negative values of the test statistic would have been considered incompatible with the null hypothesis, then the P-value has to take this into account.

6. Declare the result statistically significant if the P-value is below some critical threshold.

Ronald Fisher used $P < 0.05$ and $P < 0.01$ as convenient critical thresholds for indicating significance, and produced tables of the critical values of test statistics needed to achieve these levels of significance. The popularity of these tables led to 0.05 and 0.01 becoming established conventions, although it is now recommended that exact P-values should be reported. And it is important to emphasize that the exact P-value is conditional not only on the truth of the null hypothesis, but also on all other assumptions underlying the statistical model, such as lack of systematic bias, independent observations, and so on.

This whole process has become known as Null Hypothesis Significance Testing (NHST) and, as we shall see below, it has become a source of major controversy. But first we should examine how Fisher's ideas are used in practice.

Using Probability Theory

Perhaps the most challenging component in null-hypothesis significance testing is Step 3 – establishing the distribution of the chosen test statistic under the null hypothesis. We can always fall back on computer-intensive simulation methods as in the permutation test for the arm-crossing data, but it is far more convenient if we can use probability theory to work out the tail areas of test statistics directly, as Arbuthnot did in a simple case, and Fisher did with the hypergeometric distribution.

Often we make use of approximations that were developed by the pioneers of statistical inference. For example, around 1900 Karl Pearson developed a series of statistics for testing associations in cross-tabulations such as Table 10.1, out of which grew the classic **chi-squared test of association.***

These test statistics involve calculating the expected number of events in each cell of the table, were the null hypothesis of no-association true, and then a chi-squared statistic measures the total discrepancy between the observed and expected counts. Table 10.2 shows the expected numbers in the cells of the table, assuming the null hypothesis: for example, the expected number of females with left arm

---

* 'Chi' (pronounced 'kai') is the Greek letter χ.

	Female	Male	Total
Left arm on top	5 (5.7)	17 (16.3)	22
Right arm on top	9 (8.3)	23 (23.7)	32
Total	14	40	54

Table 10.2
Observed and expected (in parentheses) counts of arm-crossing by gender: expected counts are calculated under the null hypothesis that arm-crossing is not associated with gender.

on top is the total number of females (14), times the overall proportion of left-armers (22/54), which comes to 5.7.

It is clear from Table 10.2 that the observed and expected counts are fairly similar, reflecting that the data are just about what we would expect under the null hypothesis. The chi-squared statistic is an overall measure of the dissimilarity between the observed and expected counts (its formula is given in the Glossary), and has the value 0.02. The P-value corresponding to this statistic, available from standard software, is 0.90, showing no evidence against the null hypothesis. It is reassuring that this P-value is essentially the same as the 'exact' test based on the hypergeometric distribution.

The development and use of test statistics and P-values has traditionally formed much of a standard statistics course, and has unfortunately given the field a reputation for being largely about picking the right formula and using the right tables. Although this book tries to take a broader perspective on the subject, it is nevertheless valuable to revisit the examples we have discussed throughout the book with regard to their statistical significance.

> 1. Do the daily number of homicides in the UK follow a Poisson distribution?

Figure 8.5 showed, for England and Wales between 2014 and 2016, the observed counts of days with different numbers of homicides. There were a total of 1,545 incidents over 1,095 days, an average of 1.41 per day, and under the null hypothesis of a Poisson distribution with this mean, we would expect the counts shown in the final column of Table 10.3. Adapting the approach used for the analysis in Table 10.2, the discrepancy between the observed and expected counts can be summarized by a **chi-squared goodness-of-fit test** statistic – again, see the Glossary for details.

The observed P-value of 0.96 is not significant, so there is no evidence to reject the null hypothesis (in fact the fit is so good as to almost be suspicious). Of course we should not then assume the null hypothesis is precisely true, but it should be reasonable to use it as an assumption when assessing, for example, the changes in homicide rates seen in Chapter 9.

2. Has the unemployment rate in the UK changed in the recent past?

In Chapter 7 we saw a quarterly change in unemployment of 3,000 had a margin of error of $\pm$ 77,000, based on $\pm$ 2 standard errors. This means the 95% confidence interval runs from −80,000 to +74,000 and clearly contains the value 0, corresponding to no change in unemployment. But the fact that this 95% interval includes 0 is logically equivalent to the point estimate (−3,000) being less than 2 standard errors from 0, meaning the change is not significantly different from 0.

This reveals the essential identity between hypothesis testing and confidence intervals:

- A two-sided P-value is less than 0.05 if the 95% confidence interval does not include the null hypothesis (generally 0).

Number of homicide incidents per day	Observed days	Expected days under null hypothesis
0	259	267.1
1	387	376.8
2	261	265.9
3	131	125.0
4	40	44.1
5	13	12.4
6 or more	3	3.6
Total	1,095	1,095

Table 10.3
Observed and expected days with specified number of homicide incidents in England and Wales, April 2014 to March 2016. A chi-squared goodness-of-fit test has a P-value of 0.96, indicating no evidence against the null hypothesis of a Poisson distribution.

- A 95% confidence interval is the set of null hypotheses that are not rejected at P < 0.05.

This intimate link between hypothesis testing and confidence intervals should stop people misinterpreting results that are not statistically significantly different from 0 – this does not mean that the null hypothesis is actually true, but simply that a confidence interval for the true value includes 0. Unfortunately, as we shall see later, this lesson is often ignored.

> 3. Does taking statins reduce the risk of heart attacks and strokes in people like me?

Table 10.4 repeats the results from the Heart Protection Study (HPS) previously shown in Table 4.1, but adds columns showing the confidence with which the benefits have been established. There is a close connection between the standard errors, the confidence intervals and the P-values. The confidence intervals for the risk reduction are roughly the estimate ± 2 standard errors (note the HPS rounds the relative reductions to whole numbers). The confidence intervals easily exclude the null hypothesis of 0%, corresponding to no effect of the statin, and so the P-values are very small – in fact the P-value for the 27% reduction in heart attacks is around 1 in 3 million. This is the consequence of carrying out such a massive study.

Other summary statistics might be used, such as the difference in absolute risks, but should all give similar P-values. The HPS researchers focus on the proportional reduction since it is fairly constant across subgroups, and therefore

Event	Percentage in 10,269 people allocated statin	Percentage in 10,267 people allocated placebo	% (relative) risk reduction in those allocated statins	Standard error of risk reduction	Confidence Interval for % reduction	P-value
Heart attack	8.7	11.8	27%	4%	21% to 33%	P < 0.0001
Stroke	4.3	5.7	25%	5%	15% to 34%	P < 0.0001
Death from any cause	12.9	14.7	13%	4%	6% to 19%	P = 0.0003

Table 10.4
The results reported at the end of the Heart Protection Study, showing the estimated relative effects, their standard errors, confidence intervals and P-values testing the null hypothesis of 'no effect'.

makes a good single summary measure. There are a number of different ways of calculating the confidence intervals, although these should only produce minor differences.

> 4. Are mothers' heights associated with their sons' heights, once the fathers' heights are taken into account?

In Chapter 5 we demonstrated a multiple linear regression with son's height as the response (dependent) variable, and mother's and father's height as explanatory (independent) variables. The coefficients were shown in Table 5.3, but without any consideration of whether they could be considered significantly different from 0. To illustrate the way these results appear in statistical software, Table 10.5 reproduces the form of the output from the popular (free) R program.

As in Table 5.3, the intercept is the average of the sons' heights, and the coefficients (labelled 'Estimates' in the output) represent the expected change in height per one inch difference of their mother and father from the average mother and father heights. The standard error is calculated from a known formula, and is clearly small relative to the size of the coefficients.

The t-value, also known as a **t-statistic**, is a major focus of attention, since it is the link that tells us whether the association between an explanatory variable and the response is statistically significant. The t-value is a special case of what is known as a Student's t-statistic. 'Student' was the pseudonym of William Gosset, who developed the method in 1908

	Estimate	Std Error	t-value	Pr(>|t|)
(Intercept)	69.22882	0.10664	649.168	< 2 e–16 ***
mother's height	0.33355	0.04600	7.252	1.74 e–12 ***
father's height	0.41175	0.04668	8.820	< 2 e–16 ***

Significant codes: *** = 0.001   ** = 0.01   * = 0.05

Table 10.5
A reproduction of the output in R of a multiple regression using Galton's data, with son's height as the response variable, and mother's and father's height as explanatory variables. The t-value is the estimate divided by the standard error. The column headed Pr(>\|t\|) represents a two-sided P-value; the probability of getting such a large t-value, either positive or negative, under the null hypothesis that the true relationship is 0. The notation '2 e-16' means the P-value is less than 0.0000000000000002 (that is 15 zeros).The final line shows the interpretations of the stars in terms of P-values.

while on secondment at University College London from the Guinness brewery in Dublin – they wanted to preserve their employee's anonymity. The t-value is simply the estimate/standard error (this can be checked for the numbers in Table 10.5), and so can be interpreted as how far the estimate is away from 0, measured in the number of standard errors. Given a t-value and the sample size, the software can provide a precise P-value; for large samples, t-values greater than 2 or less than -2 correspond to P < 0.05, although these thresholds will be larger for smaller sample sizes. R uses a simple star system for P-values, from one * indicating P < 0.05, up to three stars *** indicating P < 0.001. In Table 10.5 the t-values are so large that the P-values are vanishingly small.

In Chapter 6 we saw that an algorithm might win a prediction competition by a very small margin. When predicting the survival of the *Titanic* test set, for example, the simple classification tree achieved the best Brier score (average mean squared prediction error) of 0.139, only slightly lower than the score of 0.142 from the averaged neural network (see Table 6.4). It is reasonable to ask whether this small winning margin of -0.003 is statistically significant, in the sense of whether or not it could be explained by chance variation.

This is straightforward to check, and the t-statistic turns out to be -0.54, with a two-sided P-value of 0.59.† So there is

---

† The trick is to calculate, for each of the 412 individuals in the test set, the difference between the squared prediction errors for the two algorithms; this set of 412 differences has a mean of -0.0027 and a standard deviation of 0.1028. The standard error of the estimate of the 'true' difference therefore is $0.1028/\sqrt{412} = 0.0050$, and the t-statistic is estimate / standard error $= -0.0027 / 0.0050 = -0.54$. This is known as a paired t-test since it is based on the set of differences between pairs of numbers.

no good evidence that the classification tree is truly the best algorithm! This type of analysis is not routine in Kaggle-type competitions, but it seems important to know that the winning status depends on the chance selection of cases in the test set.

Researchers spend their lives scrutinizing the type of computer output shown in Table 10.5, hoping to see the twinkling stars indicating a significant result which they can then feature in their next scientific paper. But, as we now see, this sort of obsessive searching for statistical significance can easily lead to delusions of discovery.

The Danger of Carrying Out Many Significance Tests

The standard thresholds for declaring 'significance', $P < 0.05$ and $P < 0.01$, were fairly arbitrary choices by Ronald Fisher for his tables, back in the days when calculating exact P-values was not possible using the mechanical and electrical calculators available. But what happens when we run many significance tests, each time looking to see if our P-value is less than 0.05?

Suppose a drug truly does not work; that the null hypothesis is true. If we do one clinical trial, we will declare the result as statistically significant if the P-value is less than 0.05 and, since the drug is ineffective, the chance of this happening is 0.05 or 5% – that is the definition of a P-value. This would be considered a **false-positive** result, since we incorrectly believe the drug is effective. If we do two trials, and look at the most extreme, the chance of getting at least one significant – and hence false-positive – result is close to 0.10 or 10%.[*] The chance of getting at least one false-positive result increases quickly as we do more trials; if we do ten trials of useless drugs the chance of getting at least one significant at $P < 0.05$ gets as high as 40%. This is known as the problem of **multiple testing**, and occurs whenever many significance tests are carried out and then the most significant result is reported.

A particular problem occurs when researchers split data up into many subsets, do a hypothesis test in each, and then look at the most significant. A classic demonstration was an experiment carried out by reputable researchers in 2009 which involved showing a subject a series of photographs of humans expressing different emotions, and carrying out brain imaging (fMRI) to see which regions of the subject's brain showed a significant response, taking $P < 0.001$.

The twist was that the 'subject' was a 4lb Atlantic salmon, which 'was not alive at the time of scanning'. Out of a total of 8,064 sites in the brain of this large dead fish, 16 showed a statistically significant response to the photographs. Rather than concluding the dead salmon had miraculous skills, the team correctly identified the problem of multiple testing – over 8,000 significance tests are bound to lead to false-positive results.[4] Even using a stringent criterion of $P < 0.001$, we would expect 8 significant results by chance alone.

---

[*] The exact chance of at least one trial being significant is $1 - ($ probability that both are non-significant $) = 1 - 0.95 \times 0.95 = 0.0975$, which rounds to 0.10.

One way around this problem is to demand a very low P-value at which significance is declared, and the simplest method, known as the **Bonferroni correction**, is to use a threshold of $0.05/n$, where n is number of tests done. So, for example, the tests at each site of the salmon's brain could be carried out demanding a P-value of $0.05/8,000 = 0.00000625$, or 1 in 160,000. This technique has become standard practice when searching the human genome for sites with association with diseases: since there are roughly 1,000,000 sites for genes, a P-value smaller than $0.05/1,000,000 = 1$ in 20 million is routinely demanded before claiming a discovery.

So when large numbers of hypotheses are being tested at the same time, as in brain imaging or genomics, the Bonferroni method can be used to decide whether the most extreme findings are significant. Simple techniques have also been developed that slightly relax the Bonferroni criterion for the second most extreme result, the third most extreme and so on, that are designed to control the overall proportion of 'discoveries' that turn out to be false claims – the so-called **false discovery rate**.

Another way to avoid false-positives is to demand replication of the original study, with the repeat experiment carried out in entirely different circumstances, but with essentially the same protocol. For new pharmaceuticals to be approved by the US Food and Drug Administration, it has become standard that two independent clinical trials must have been carried out, each showing clinical benefit that is significant at $P < 0.05$. This means that the overall chance of approving a drug, that in truth has no benefit at all, is $0.05 \times 0.05 = 0.0025$, or 1 in 400.

5. Does the Higgs boson exist?

Throughout the twentieth century, physicists developed a 'standard model' intended to explain the forces operating at a subatomic level. One piece of the model remained an unproved theory: the 'Higgs field' of energy which permeates the universe, and gives mass to particles such as electrons through its own fundamental particle, the so-called Higgs boson. When researchers at CERN finally reported the discovery of the Higgs boson in 2012, it was announced as a 'five-sigma' result.[5] But few people would have realized this was an expression of statistical significance.

When the researchers plotted the rate at which specific events occurred for different energy levels, the curve was found to have a distinct 'hump' just where it would be expected if the Higgs boson existed. Crucially, a form of chi-squared goodness-of-fit test revealed a P-value of less than 1 in 3.5 million, under the null hypothesis that the Higgs did *not* exist and the 'hump' was simply the result of random variation. But why was this reported as a 'five-sigma' discovery?

It is standard in theoretical physics to report claims of discoveries in terms of 'sigmas', where a 'two-sigma' result is an observation that is two standard errors away from the null hypothesis (remember that we used sigma (σ) as the Greek letter representing a population standard deviation): the 'sigmas' in theoretical physics correspond precisely to the *t*-value in the computer output shown in Table 10.5 for the multiple regression example. Since an observation that gave a two-sided P-value of 1 in 3.5 million – that observed from the chi-squared test – would be five standard errors from the null hypothesis, the Higgs boson was therefore said to be a five-sigma result.

The team at CERN clearly did not want to announce their 'discovery' until the P-value was extremely small. First, they needed to allow for the fact that significance tests had been carried out at all energy levels, not just the one in the final chi-squared test – this adjustment for multiple testing is known as the 'look elsewhere effect' in physics. But mainly they wanted to be confident that any attempt at replication would come up with the same conclusion. It would simply be too embarrassing to make an incorrect claim about the laws of physics.

To answer the question at the start of this section: it seems reasonable now to assume that the Higgs boson exists. This becomes the new null hypothesis until, perhaps, a deeper theory is suggested.

Neyman–Pearson Theory

Why did the Heart Protection Study need over 20,000 participants?

The Heart Protection Study was huge, but its size was not some arbitrary choice. When planning the trial, the researchers had to say how many people would need to be randomized to statins or not, and this had to have a strong statistical foundation in order to justify the expense of such an experiment. Their plan was based on statistical ideas that were developed by Jerzy Neyman and Egon Pearson, whom we met previously as the developers of confidence intervals.

The idea of P-values and significance testing was developed by Ronald Fisher in the 1920s as a way of checking the adequacy of a specified hypothesis. If a small P-value is observed, then either something very surprising has happened, or the null hypothesis is untrue: the smaller the P-value, the more evidence that the null hypothesis might be an inappropriate assumption. This was intended as a fairly informal procedure, but in the 1930s Neyman and Pearson developed a theory of inductive behaviour which attempted to put hypothesis testing on a more rigorous mathematical footing.

Their framework required specification of not only a null hypothesis, but also an alternative hypothesis which represents a more complex explanation for the data. They then considered the possible decisions after a hypothesis test, which are either to reject a null hypothesis in favour of the alternative, or not to reject the null.* Two types of mistake are therefore possible: a **Type I error** is made when we reject a null hypothesis when it is true, and a **Type II error** is made when we do not reject a null hypothesis when in fact the alternative hypothesis holds. There is a strong legal analogy which is illustrated in Table 10.6 – a Type I legal error is to falsely convict an innocent person, and a Type II error is to find someone 'not guilty' when in fact they did commit the crime.

When planning an experiment, Neyman and Pearson suggested that we should choose two quantities which together

---

* Neyman and Pearson's original theory included the idea of 'accepting' a null hypothesis, but this part of their theory is now ignored.

Truth	Outcome of hypothesis test	
	Do not reject null hypothesis (find suspect 'not guilty')	Reject null hypothesis in favour of alternative (find suspect guilty)
Null hypothesis (suspect is innocent)	Correct in not rejecting null. Correctly find an innocent person 'not guilty'.	Type I error: Incorrectly reject null. Wrongly convict an innocent person
Alternative hypothesis (suspect is guilty)	Type II error: Incorrectly fail to reject null. Wrongly fail to convict a guilty person	Correctly reject null. Correctly convict a guilty person

Table 10.6
Possible outcomes of a hypothesis test, with the analogy of a criminal trial.

will determine how large the experiment should be. First, we should fix the probability of a Type I error, given the null is true, at a pre-specified value, say 0.05; this is known as the **size of a test**, and generally denoted α (alpha). Second, we should pre-specify the probability of a Type II error, given the alternative hypothesis is true, generally known as β (beta). In fact researchers generally work in terms of $1 - \beta$, which is termed the **power of a test**, and is the chance of rejecting the null in favour of an alternative hypothesis, given the latter is true. In other words, the power of an experiment is the chance that it will correctly detect a real effect.

There is a close connection between the size α and Fisher's P-value. If we take α as the threshold at which we consider results significant, then the results that lead us to reject the null will be exactly those for which P less than α. So α can be considered as the threshold significance level – an α of 0.05 means that we reject the null for all P-values less than 0.05.

Formulae exist for the size and power of different forms of experiment, and they each depend crucially on sample size. But if the sample size is fixed, there is an inevitable trade-off: to increase power, we can always make the threshold for 'significance' less stringent and so make it more likely we will correctly identify a true effect, but this means increasing the chance of a Type I error (the size). In the legal analogy, we can loosen the criteria for conviction, say by loosening the requirement of proof 'beyond reasonable doubt', and this will result in more criminals being correctly convicted, but at the inevitable cost of more innocent people being incorrectly found guilty.

The Neyman–Pearson theory had its roots in industrial quality control, but is now used extensively in testing new medical treatments. Before starting a randomized clinical trial, the protocol will specify a null hypothesis that the treatment has no effect, and an alternative hypothesis, generally an effect that is considered both plausible and important. The researchers then lay down the size and power of the study, often setting α = 0.05 and β = 0.80. This means they demand a P-value of less than 0.05 to declare the result significant, and have 80% chance of this being achieved if the treatment is truly effective: together these give rise to an estimate of the number of participants that are needed.

Researchers need to be more stringent if they want to carry out a definitive clinical trial. For example, the Health Protection Study concluded that

> if cholesterol-lowering therapy reduced 5-year coronary heart disease mortality by about 25% and all-cause mortality by about 15%, then a study of this size with good compliance would have an excellent chance of demonstrating such effects at convincing levels of statistical significance (i.e., >90% power to achieve p<0.01).

In other words, if the true treatment effect is a 25% reduction in heart disease mortality and 15% all-cause mortality (the alternative hypotheses), the study has around β = 90% power and α = 1% size. These requirements dictated a sample size of over 20,000. In fact, as Table 10.4 shows, the final results included a 13% reduction in all-cause mortality, remarkably close to that which had been planned for.

The idea of having a large enough sample to have sufficient power to detect a plausible alternative hypothesis has become totally entrenched in planning medical studies. But studies in psychology and neuroscience often have sample sizes chosen on the basis of convenience or tradition, and can be as low as 20 subjects per condition being studied. True, and interesting, alternative hypotheses may be missed through studies simply being too small, and the need for other experimental areas to think about the power of their experiments is finally being recognized.

As we shall see in the next chapter, Neyman and Pearson had vehement, even publicly abusive, arguments with Fisher over the appropriate form of hypothesis testing, and this conflict has never been resolved into a single 'correct' approach. The Heart Protection Study shows that clinical trials tend to be designed from a Neyman–Pearson perspective but, strictly speaking, size and power are irrelevant once the experiment has actually been carried out. At this point the trial is analysed using confidence intervals to show the plausible values for the treatment effects, and Fisherian P-values to summarize the strength of the evidence against the null hypothesis. So an odd mixture of Fisher's and Neyman–Pearson's ideas has proven to be remarkably effective.

Could Harold Shipman have been caught earlier?

We saw in the Introduction that Dr Harold Shipman murdered over 200 of his patients over the course of twenty years before he was finally caught. The families of his victims were naturally distressed that he could have carried out his crimes for so long without attracting suspicion, and a subsequent public inquiry was tasked with judging if he could have been identified earlier. In advance of the inquiry, the number of death certificates signed by Shipman for people dying in their homes or in his practice since 1977 had been accumulated, and compared with the number that would have been expected, given the age composition of all the patients under Shipman's 'care' and the mortality rates for other GPs in the surrounding area. Making this sort of comparison means that local conditions such as changing temperature and flu outbreaks are controlled for. Figure 10.3 shows the results obtained by subtracting the expected number from the observed number of death certificates, accumulated from 1977 until Shipman's arrest in 1998. This difference can be termed his 'excess' mortality.

By 1998, his estimated excess mortality for people aged 65 or over was 174 women and 49 men. This was almost exactly the number of older people later confirmed to be victims by the Inquiry, showing the remarkable accuracy of this purely statistical analysis, for which no knowledge of individual cases had been included.[6]

Supposing, in some fictitious history, that someone had been monitoring Shipman's deaths year by year and doing the calculations necessary to produce Figure 10.3; at what point might they have 'blown the whistle'? They could, for example, have performed a significance test at the end of each year. Counts of deaths, like homicides, are the result of many individuals having a small probability of the event, and can be assumed to have a Poisson distribution, and so the null hypothesis would be that the cumulative observed numbers

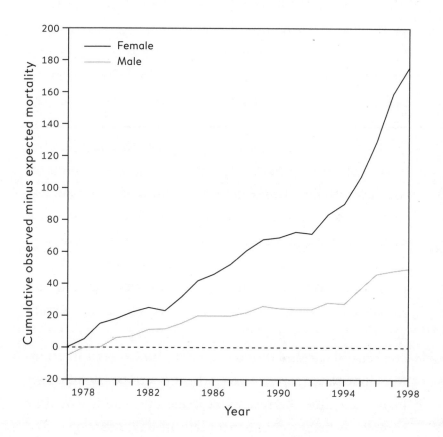

Figure 10.3
Cumulative number of death certificates signed by Shipman for patients who were age 65 or over and who died at home or in his practice. The expected number, given the composition of his practice list, has been subtracted.

of deaths was an observation from a Poisson distribution with expectation given by the cumulative expected counts.

If this had been done using the total deaths for men and women shown in Figure 10.3, then in 1979, after only three years of monitoring, there would have been a one-sided P-value of 0.004 arising from comparing 40 observed deaths while only expecting 25.3.* The results could have been declared 'statistically significant' and Shipman investigated and detected.

But there are two reasons why such a statistical procedure would have been grossly inappropriate as a way of monitoring the mortality rates of general practitioners. First, unless there was some other reason to suspect Shipman and to set up a monitoring process for him alone, we would have been calculating such P-values for all the general practitioners in the UK – numbering around 25,000 at the time. As seen with the dead salmon, we know that if we carry out enough significance tests we will get false signals. With 25,000 GPs tested at a critical threshold of 0.05, we would expect 1 in 20 utterly innocent doctors – around 1,300 – to be 'significantly high' each time the test was carried out, and it would be completely inappropriate to investigate all these people. And Shipman might be lost in all these false-positives.

An alternative is to apply the Bonferroni method, and demand a P-value of 0.05/25,000, or 1 in 500,000, for the most extreme GP; for Shipman this would have occurred

---

* The P-value is one-sided since we are only interested in detecting increased mortality, and not decreased. The P-value is therefore the probability that a Poisson random variable with mean 22.5 would be at least 40, which from standard software is 0.004.

in 1984, when he had 105 deaths compared to the 59.2 that would be expected, an excess of 46.

But even this would not be a reliable procedure to apply to all the GPs in the country. For the second problem is that we are carrying out repeated significance tests, as each year's new data are added on and another test performed. There is some remarkable but complex theory, known by the delightful term 'the Law of the Iterated Logarithm', that shows that if we carry out such repeated testing, even if the null hypothesis is true, then we are *certain* to eventually reject that null at any significance level we choose.

This is very worrying, as it means that if we keep testing a doctor for long enough then we are guaranteed to eventually think we have found evidence of excess mortality, even if in reality their patients are not subject to any excess risk. Fortunately there are statistical methods for dealing with this problem of **sequential testing**, first developed in the Second World War by teams of statisticians who had nothing to do with medical care, but were working on industrial quality control of armaments and other war material.

Items coming off the production line were being tested for conformity to a standard, and the whole process was monitored by steadily accumulating total deviations from the standard, much in the same way as monitoring excess mortality. These scientists realized that the Law of the Iterated Logarithm meant that repeated significance testing would always lead eventually to an alert that the industrial process had gone out of strict control, even if in truth everything was functioning fine. Statisticians in the US and UK, working independently, developed what became known as the Sequential Probability Ratio Test (SPRT), which is a statistic that monitors accumulating evidence about deviations, and can at any time be compared with simple thresholds – as soon as one of these thresholds is crossed, then an alert is triggered and the production line is investigated.* Such techniques led to more efficient industrial processes, and were later adapted for use in so-called sequential clinical trials in which accumulated results are repeatedly monitored to see if a threshold that indicates a beneficial treatment has been crossed.

I was one of a team that developed a version of the SPRT that could be applied to the Shipman data. This is plotted in Figure 10.4 for both men and women, assuming an alternative hypothesis that Shipman had double the mortality rate of his colleagues. The test has thresholds that control the Type I (alpha) and Type II (beta) error probability to the specified values of 1 in 100, 1 in 10,000 and 1 in 1,000,000: the Type I error is the overall probability of the test statistic crossing the threshold at some point, given that Shipman had the expected mortality rates, and the Type II error is the overall probability of the test statistic *not* crossing the threshold at some point, given that Shipman had double the expected mortality rate.[7]

Given there are around 25,000 GPs, then a threshold P-value of 0.05/25,000 or 1 in 500,000 might be reasonable. For women alone, Shipman would have crossed the more

---

* The statisticians were led by Abraham Wald in the US and George Barnard in the UK. Barnard was a delightful man, a pure mathematician (and Communist) before the war, when like many others he adapted his skills for statistical war work. He later went on to develop the official British Standard for condoms (BS 3704).

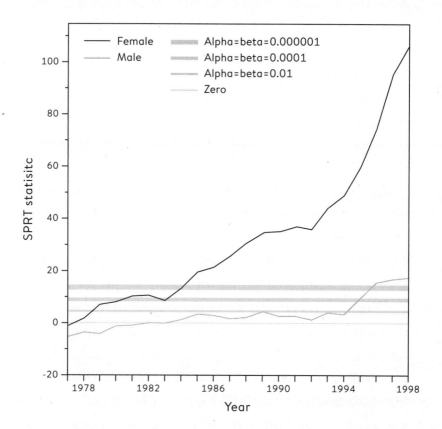

Figure 10.4
Sequential Probability Ratio Test (SPRT) statistic for detection of
a doubling in mortality risk: patients aged >64 and dying in home/
practice. The straight lines indicate thresholds for 'alerting', which
provide the overall Type 1 (alpha) and Type II (beta) error rates that
are shown – these are assumed to be equal. Looking at the line for
females, it is apparent that Shipman would have crossed the outer
threshold in 1985.

stringent alpha = 0.000001, or 1 in a million, threshold in 1985, and if men and women are combined, he would have done so in 1984. So in this case the correct sequential test would have sounded the alarm at the same time as the naïve repeated significance test.

Our conclusion to the public inquiry was that if someone had been doing this monitoring, and Shipman had been investigated in 1984 and prosecuted, then around 175 lives could have been saved. All by a routine application of a simple statistical monitoring procedure.

A monitoring system for general practitioners was subsequently piloted, which immediately identified a GP with even higher mortality rates than Shipman! Investigation revealed this doctor practised in a south-coast town with a large number of retirement homes with many old people, and he conscientiously helped many of his patients to remain out of hospital for their death. It would have been completely inappropriate for this GP to receive any publicity for his apparently high rate of signing death certificates. The lesson here is that while statistical systems can detect outlying outcomes, they cannot offer reasons why these might have occurred, so they need careful implementation in order to avoid false accusations. Another reason to be cautious about algorithms.

What could go wrong with P-values?

Ronald Fisher developed the idea of a P-value as a measure of the compatibility of the data with some preformed hypothesis. So if you calculate a P-value and find it is small, it means that it is unlikely that your summary statistic would be so extreme if the hypothesis were true, and so either something surprising has occurred or your original hypothesis is faulty. The logic may be convoluted but we have seen how useful this basic idea can be. So what could possibly go wrong?

Rather a lot, it turns out. Fisher envisaged the sort of situation seen in the early examples in this chapter, with a single set of data, a single summary outcome measure and a single test of compatibility. But in the last few decades P-values have become the currency of research, with vast numbers appearing in the scientific literature – a study scraped around 30,000 t-statistics and their accompanying P-values from just three years of papers in eighteen psychology and neuro-science journals.[8]

So let's see what we would expect to happen with, say, 1,000 studies, each designed with size 5% (α) and 80% power ($1-\beta$), noting that in practice most studies would have considerably less than 80% power. In the real world of research, although experiments are carried out in the hope of making a discovery, it is recognized that most null hypotheses are (at least approximately) true. So suppose only 10% of the null hypotheses tested were actually false: even this figure is probably rather high for new pharmaceuticals, which have a notoriously low success rate. Then, in a similar way to the screening examples in Chapter 8, Figure 10.5 shows the frequencies for what we expect to happen for these 1,000 studies.

This reveals that we expect to claim 125 'discoveries', but of these 45 are false-positives: in other words 36%, or over a third, of the rejected null hypotheses (the 'discoveries') are

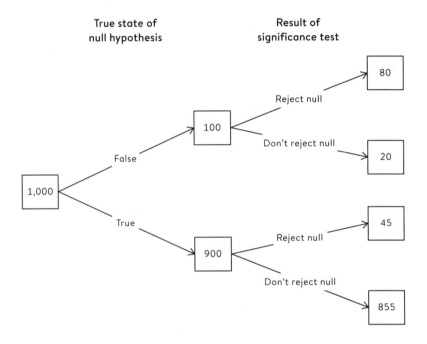

Figure 10.5

The expected frequencies of the outcomes of 1,000 hypothesis tests carried out with size 5% (Type I error, α) and 80% power (1 – Type II error, 1–β). Only 10% (100) of the null hypotheses are false, and we correctly detect 80% of them (80). Of the 900 null hypotheses that are true, we incorrectly reject 45 (5%). Overall, of 125 'discoveries', 36% (45) are false discoveries.

incorrect claims. This rather gloomy picture becomes even worse when we consider what actually ends up in the scientific literature, since journals are biased towards publishing positive results. A similar analysis of scientific studies led to Stanford professor of medicine and statistics John Ioannidis's famous claim in 2005 that 'most published research findings are false'.[9] We shall return to the reasons for his dismal conclusion in Chapter 12.

Since all these false discoveries were based on a P-value identifying a 'significant' result, P-values have been increasingly blamed for a flood of incorrect scientific conclusions. In 2015 a reputable psychology journal even announced that they would ban the use of NHST (Null Hypothesis Significance Testing). Finally in 2016 the American Statistical Association (ASA) managed to get a group of statisticians to agree on six principles about P-values.*

The first of these principles simply points out what P-values can do:

1. P-values can indicate how incompatible the data are with a specified statistical model.

As we have repeatedly seen, P-values do this by essentially measuring how surprising the data are, given a null hypothesis that something does not exist. For example, we ask whether the data are incompatible with a drug that doesn't work? The logic can be tricky, but useful.

---

* This was a remarkable achievement, given the collective noun for statisticians has been said to be a 'variance'.

The second principle tries to remedy errors in their interpretation:

2. P-values do not measure the probability that the studied hypothesis is true, or the probability that the data were produced by random chance alone.

Back in Chapter 8, we were very careful to distinguish appropriate conditional probability statements such as 'only 10% of women without breast cancer would get a positive mammogram' from the incorrect 'only 10% of women with a positive mammogram do not have breast cancer'. This was the mistake known as the prosecutor's fallacy, and we saw there are neat ways of remedying this error by thinking of what we might expect to happen to 1,000 women being tested.

Similar problems can occur with P-values, which measure the chance of such extreme data occurring, if the null hypothesis is true, and do *not* measure the chance that the null hypothesis is true, given that such extreme data have occurred. This is a subtle but essential difference.

When the CERN teams reported a 'five-sigma' result for the Higgs boson, corresponding to a P-value of around 1 in 3.5 million, the BBC reported the conclusion correctly, saying this meant 'about a one-in-3.5 million chance that the signal they see would appear if there were no Higgs particle.' But nearly every other outlet got the meaning of this P-value wrong. For example, *Forbes Magazine* reported, 'The chances are less than 1 in a million that it is not the Higgs boson', a clear example of the prosecutor's fallacy. *The Independent* was typical in claiming that 'there is less than a one in a million chance that their results are a statistical fluke'. This may not be blatantly mistaken as *Forbes*, but it is still assigning the small probability to 'their results are a statistical fluke', which is logically the same as saying this is the probability of the null hypothesis being tested. That's why the ASA try to emphasize that the P-value is *not* 'the probability that the data were produced by random chance alone'.

The ASA's third principle seeks to counter the obsession with statistical significance:

3. Scientific conclusions and business or policy decisions should not be based only on whether a P-value passes a specific threshold.

When Ronald Fisher started publishing tables showing values of statistics that would just make the results 'P < 0.05' or 'P < 0.01', he presumably had little idea of how such rather arbitrary thresholds would come to dominate scientific publications, with all results tending to be separated into 'significant' or 'not significant'. From there it is a short step to consider 'significant' results as proven discoveries, producing an oversimplified and dangerous precedent for going from data straight to conclusions without pausing for thought on the way.

A dire consequence of this simple dichotomy is the misinterpretation of 'not significant'. A non-significant P-value suggests the data are compatible with the null hypothesis, but this does not mean the null hypothesis is precisely true. After all, just because there's no direct evidence that a criminal was at the scene of a crime, that does not mean he is innocent. But this mistake is surprisingly common.

Consider the major scientific dispute about whether a

small amount of alcohol, say one drink a day, is good for you. A study claimed that only older women might benefit from moderate alcohol consumption, but close inspection revealed that other groups also showed a benefit, but that it was not statistically significant, since the confidence intervals around the estimated benefit in these groups were very wide indeed. Although the confidence intervals included 0 and hence the effects were not statistically significant, the data were fully compatible with the 10% to 20% reduction in mortality risk that had been previously suggested. But *The Times* trumpeted that 'Alcohol Has No Health Benefits After All'.[10]

To summarize, it is very misleading to interpret 'not significantly different from 0' as meaning that the true effect actually *was* 0, particularly in smaller studies with low power and wide confidence intervals.

The ASA's fourth principle sounds fairly innocuous:

4. Proper inference requires full reporting and transparency.

The most obvious need is to clearly report how many tests were actually done, so if the most significant result is being emphasized, we can apply some form of adjustment such as Bonferroni. But the problems with selective reporting can be much more subtle than this, as we shall see in the next chapter. Only by knowing the plan of the study, and what was actually done, can problems with P-values be avoided.

You have planned your study, collected your data, done your analysis and got a 'significant' result. So surely this must be an important discovery? The ASA's fifth principle warns you not to be too arrogant:

5. A P-value, or statistical significance, does not measure the size of an effect or the importance of a result.

Our next example shows that, particularly if we have large samples, we may be reasonably confident that an association exists but still be distinctly unimpressed by its importance.

'Why Going to University Increases Risk of Getting a Brain Tumour'

We saw this headline in Chapter 4. After adjusting for marital status and income in a regression analysis, the Swedish researchers found a 19% relative increase in risk between the lowest education level (just primary) and the highest (higher university degree), with a 95% confidence interval of 7% to 33% – interestingly, the paper did not report any P-values, but since the 95% interval for the relative risk excludes 1, we can conclude that $P < 0.05$.

By now the reader should be ready with a list of potential concerns about the conclusions, but the authors pre-empted these. Alongside the results of their study they acknowledged that:

- no causal interpretation could be made;
- no adjustment had been made for potential lifestyle confounders such as alcohol consumption;
- people of higher economic status are likely to have a greater tendency to seek care, and hence there could be reporting bias.

But one important feature was not mentioned: the small size of the apparent association. A 19% increase between the lowest and highest educational levels is much lower than is found for many cancers. The paper reported that 3,715 brain tumours were diagnosed in over 2,000,000 men over 18 years (roughly 1 in 600) and so, following the procedure outlined in Chapter 1 to translate relative risks into changes in absolute risk, we can calculate that:

- out of around 3,000 men of the lowest educational level, we would expect around 5 tumours to be diagnosed (1 in 600 baseline risk);
- out of 3,000 men of the highest educational level, we would expect 6 (a 19% relative increase).

This gives a somewhat different impression of the findings, and is in fact rather reassuring. Such a small increased risk in a rare cancer could only be found to be statistically significant when huge numbers of people are studied: in this case over two million men.

The main lessons from this scientific study might therefore be (a) that 'big data' can easily lead to findings that are statistically significant but not of **practical significance**, and (b) that you should not be concerned that studying for your degree is going to give you a brain tumour.

The final principle from the ASA is rather more subtle:

6. By itself, a P-value does not provide a good measure of evidence regarding a model or hypothesis. For example, a P-value near 0.05 taken by itself offers only weak evidence against the null hypothesis.

This claim, partly based on the 'Bayesian' reasoning outlined in the next chapter, has led a prominent group of statisticians to argue that the standard threshold for a 'discovery' of a new effect should be changed to $P < 0.005$.[11]

What effect might this have? Changing the criterion for 'significance' from 0.05 (1 in 20) to 0.005 (1 in 200) in Figure 10.5 would mean that instead of having 45 false-positive 'discoveries', we would have only 4.5. This would reduce the total number of discoveries to 84.5, and of these only 4.5 (5%) would be false discoveries. Which would be a considerable improvement from 36%.

Fisher's original idea for testing hypotheses has been of great benefit to the practice of statistics and the prevention of unjustified scientific claims. But statisticians have frequently complained about the willingness of some researchers to lurch casually from P-values taken from poorly designed studies to confident generalizable inferences: a kind of alchemy for turning uncertainty into certainty, mechanically applying statistical tests to split all results into 'significant' and 'nonsignificant'. We shall see some of the poor consequences of this behaviour in Chapter 12, but first turn to an alternative approach to statistical inference that entirely rejects the whole idea of null-hypothesis significance testing.

So, as another mind-stretching requirement of statistical science, it would help if you could (temporarily) forget everything you may have learned from this and preceding chapters.

Summary

- Tests of null hypotheses – default assumptions about statistical models – form a major part of statistical practice.
- A P-value is a measure of the incompatibility between the observed data and a null hypothesis: formally it is the probability of observing such an extreme result, were the null hypothesis true.
- Traditionally, P-value thresholds of 0.05 and 0.01 have been set to declare 'statistical significance'.
- These thresholds need to be adjusted if multiple tests are conducted, for example on different subsets of the data or multiple outcome measures.
- There is a precise correspondence between confidence intervals and P-values: if, say, the 95% interval excludes 0, we can reject the null hypothesis of 0 at $P < 0.05$.
- Neyman–Pearson theory specifies an alternative hypothesis, and fixes Type I and Type II error rates for the two possible kinds of errors in a hypothesis test.
- Separate forms of hypothesis tests have been developed for sequential testing.
- P-values are often misinterpreted: in particular they do not convey the probability that the null hypothesis is true, nor does a non-significant result imply that the null hypothesis is true.

CHAPTER 11

Learning from Experience the Bayesian Way

> I am not at all sure that the 'confidence' is not a 'confidence trick'.
>
> — Arthur Bowley, 1934

I must now make an admission on behalf of the statistical community. The formal basis for learning from data is a bit of a mess. Although there have been numerous attempts to produce a single unifying theory of statistical inference, none has been fully accepted. It is no wonder mathematicians tend to dislike teaching statistics.

We have already met the competing ideas of Fisher and Neyman–Pearson, and it is time to explore a third, Bayesian, approach to inference. This has only come to prominence in the last fifty years, but its basic principles go back somewhat further, in fact to the Reverend Thomas Bayes, a Nonconformist minister turned probability theorist and philosopher from Tunbridge Wells, who died in 1761.*

---

* He died with no knowledge whatsoever of his enduring legacy, and not only was his seminal paper published posthumously in 1763, but his name did not become associated with this approach until the twentieth century.

The good news is that the Bayesian approach opens fine new possibilities for making the most of complex data. The bad news is that it means putting aside almost everything you may have learned in this book and elsewhere about estimation, confidence intervals, P-values, hypothesis testing, and so on.

What Is the Bayesian Approach?

Thomas Bayes' first great contribution was to use probability as an expression of our lack of knowledge about the world or, equivalently, our ignorance about what is currently going on. He showed that probability can be used not only for future events subject to random chance – aleatory uncertainty, to use the term introduced in Chapter 8 – but also for events which are true, and might well be known to some people, but that we are not privy to – so-called epistemic uncertainty.

If you briefly think about it, we are surrounded by epistemic uncertainty about things that are fixed but unknown to us. Gamblers bet on the next card to be dealt, we buy lottery scratch-cards, we discuss the possible gender of a baby, we puzzle over whodunnits, we argue over the numbers of tigers left in the wild, and we are told estimates of the possible number of migrants or the unemployed. All these are facts or quantities that exist out there in the world, but we just do not know what they are. To emphasise again, from a Bayesian perspective, it is fine to use probabilities to represent our personal ignorance about these facts and numbers. We might even think of putting probabilities on alternative scientific theories, but this is more contested.

These probabilities will of course depend on our current knowledge: remember from Chapter 8 how our probability of whether a coin has come up heads or tails depends on whether we have looked at it or not! So these Bayesian probabilities are necessarily subjective – they depend on our relationship with the outside world, and are not properties of the world itself. These probabilities should change as we receive new information.

Which brings us to Bayes' second key contribution: a result in probability theory that allows us to continuously revise our current probabilities in the light of new evidence. This has become known as **Bayes' theorem**, and essentially provides a formal mechanism for learning from experience, which is an extraordinary achievement for an obscure clergyman from a small English spa town. Bayes' legacy is the fundamental insight that the data does not speak for itself – our external knowledge, and even our judgement, has a central role. This may seem to be incompatible with the scientific process, but of course background knowledge and understanding has always been an element in learning from data, and the difference is that in the Bayesian approach it is handled in a formal and mathematical way.

The implications of Bayes' work have been deeply contested, with many statisticians and philosophers objecting to the idea that subjective judgement has any role in statistical science. So it is only fair that I make my personal position clear: I was introduced into a 'subjectivist' Bayesian school of statistical reasoning at the start of my career,* and it still remains for me the most satisfying approach.

---

* Some might even say I was indoctrinated.

> You have three coins in your pocket: one has two
> heads, one is fair and one has two tails. You pick a coin
> at random and flip it, and it comes up heads. What
> should be your probability that the other side of the
> coin also shows heads?

This is a classic problem in epistemic uncertainty: there is no randomness left in the coin once it has been flipped, and any probability is simply an expression of your current personal ignorance about the other side of the coin.

Many people would jump to the conclusion that the answer is ½, since the coin must be either the fair or the two-headed coin, and each was equally likely to be picked. There are many ways to check whether this is correct, but the easiest is to use the idea of expected frequencies demonstrated in Chapter 8.

Figure 11.1 shows what would you expect to see if you carried out this exercise six times. On average, each coin would be chosen twice, and each side of each coin would turn up in the flip. Three of the flips end up in a head, and in two of these the coin is two-headed. So your probability that the chosen coin is two-headed rather than fair should be ⅔, and not ½. Essentially, seeing a head makes it more likely that the two-headed coin has been chosen, since this coin provides two opportunities for a head to land face-up, whereas the fair coin only provides one.

If this result seems unintuitive, then the next example might be even more surprising.

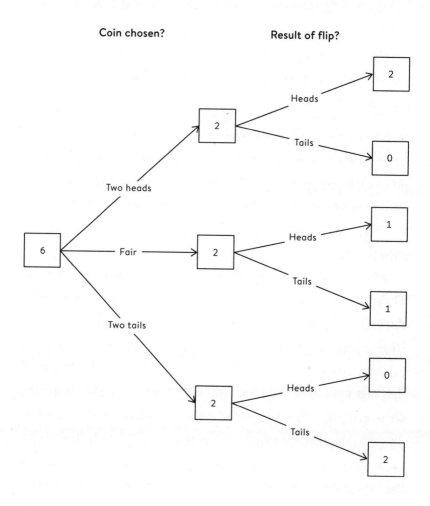

Figure 11.1
Expected frequency tree for three-coin problem, showing what we would expect to happen in six repetitions.

> Suppose a screening test for doping in sports is claimed to be '95% accurate', meaning that 95% of dopers, and 95% of non-dopers, will be correctly classified. Assume 1 in 50 athletes are truly doping at any time. If an athlete tests positive, what is the probability that they are truly doping?

This type of potentially challenging problem is again best dealt with using expected frequencies, similar to the analysis of breast screening in Chapter 8, and the claims in Chapter 10 that a high proportion of the published scientific literature is wrong.

The tree in Figure 11.2 starts with 1,000 athletes, of whom 20 are doping and 980 are not. All but one of them are detected (95% of 20 = 19), but 49 non-dopers also have positive tests (95% of 980 = 931). We therefore expect a total of 19 + 49 = 68 positive tests, of whom only 19 are truly doping. So if someone tests positive, there is only 19/68 = 28% chance they are truly doping – the remaining 72% of positive tests are false accusations. Even though drug testing could be claimed to be '95% accurate', the majority of people who test positive are in fact innocent – it does not require much imagination to see the problems this apparent paradox could cause in real life, with athletes being casually condemned because they failed a drug test.

One way of thinking of this process is that we are 'reversing the order' of the tree to put testing first, followed by the revelation of the truth. This is shown explicitly in Figure 11.3.

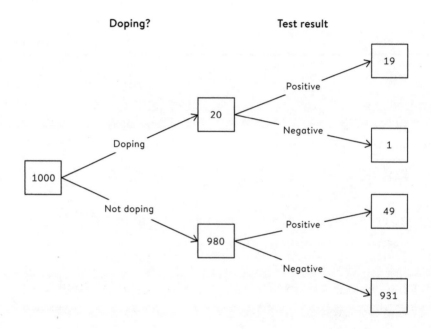

Figure 11.2
Expected frequency tree for sports doping, showing what we
expect to happen to 1,000 athletes when 1 in 50 are doping, and
the screening test is '95% accurate'.

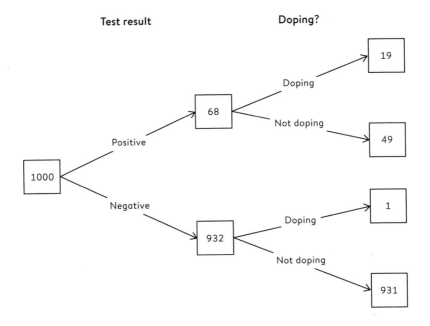

Test result   **Doping?**

19

Doping

68

Not doping

49

Positive

1000

Negative

1

Doping

932

Not doping

931

Figure 11.3
'Reversed' expected frequency tree for sports doping, restructured
so that the test result comes first, followed by revealing the true
activity of the athlete.

This 'reversed tree' arrives at exactly the same numbers for the final outcomes, but respects the temporal order in which we come to know things (testing and then the truth about doping), rather than the actual timeline of underlying causation (doping and then testing). This 'reversal' is exactly what Bayes' theorem does – in fact Bayesian thinking was known as 'inverse probability' until the 1950s.

The sports doping example shows how easy it is to confuse the probability of doping, given a positive test (28%), with the probability of testing positive, given doping (95%). We have already seen other contexts when the probability of 'A given B' is confused with the probability of 'B given A':

- the misinterpretation of P-values, in which the probability of the evidence given the null hypothesis is confused with the probability of the null hypothesis given the evidence.
- the prosecutor's fallacy in court cases, in which the probability of the evidence given innocence is confused with the probability of innocence given the evidence.

A reasonable observer might think that formal Bayesian thinking would bring clarity and rigour to the handling of evidence in legal cases, and so may be surprised to hear that Bayes' theorem is essentially prohibited from British courts. Before revealing the arguments behind this ban, we must first look at the statistical quantity that *is* allowed in court – the **likelihood ratio.**

Odds and Likelihood Ratios

The doping example lays out the logical steps necessary to get to the quantity that is really of interest when making decisions: *out of people who test positive, the proportion who are really doping*, which turns out to be 19/68. The expected frequency tree shows that this depends on three crucial numbers: the proportion of athletes who are doping (1/50, or 20/1,000 in the tree), the proportion of doping athletes who correctly test positive (95%, or 19/20 in the tree) and the proportion of non-doping athletes who incorrectly test positive (5%, or 49/980 in the tree).

The analysis becomes (fairly) intuitive by using an expected frequency tree, although Bayes' theorem can also be expressed in a convenient formula using probabilities. But first we need to return to the idea of odds introduced in Chapter 1, to which seasoned gamblers will need no introduction, at least if they are British. The odds of an event is the probability of it happening, divided by the probability of it *not* happening. So the odds of flipping a coin and getting a head is 1, which comes from ½ (the probability of a head) divided by ½ (the probability of getting a tail).* The odds of throwing a die and getting a six is 1/6 divided by 5/6, which comes to 1/5, popularly known as '1 to 5 on', or '5 to 1 against' if you use the British method of expressing gambling odds.

Next we need to introduce the idea of a likelihood ratio,

---

* Odds of 1 are sometimes known as 'evens', since the events are equally likely, or evenly balanced.

a concept that has become critical in communicating the strength of forensic evidence in criminal court cases. Judges and lawyers are being increasingly trained to understand likelihood ratios, which essentially compare the relative support provided by a piece of evidence for two competing hypotheses, which we shall call A and B, but which would often represent guilt or innocence. Technically, the likelihood ratio is the probability of the evidence assuming hypothesis A, divided by the probability of the evidence assuming hypothesis B.

Let us see how this works in the doping case, where the forensic 'evidence' is the positive test result, hypothesis A is that the athlete is guilty of doping, and hypothesis B is that they are innocent. We are assuming that 95% of dopers test positive, so the probability of the evidence, given hypothesis A, is 0.95. We know that 5% of non-dopers test positive, so the probability of the evidence, given hypothesis B, is 0.05. So the likelihood ratio is 0.95/0.05 = 19: that is, the positive test result was 19 times more likely to happen were the athlete guilty rather than innocent. This may at first seem like quite strong evidence, but we shall later come to likelihood ratios in millions and billions.

So let's put all this together in Bayes' theorem, which simply says that

the initial odds for a hypothesis × the likelihood ratio = the final odds for the hypothesis

For the doping example, the initial odds for the hypothesis 'the athlete is doping' is 1/49, and the likelihood ratio is 19, so Bayes' theorem says the final odds are given by

$$1/49 \times 19 = 19/49$$

These odds of 19/49 can be transformed to a probability of 19/(19+49) = 19/68 = 28%. So this probability, which was obtained from the expected frequency tree in a rather simple way, can also be derived from the general equation for Bayes' theorem.

In more technical language, the initial odds are known as the 'prior' odds, and the final odds are the 'posterior' odds. This formula can be repeatedly applied, with the posterior odds becoming the prior odds when introducing new, independent items of evidence. When combining all the evidence, this process is equivalent to multiplying the independent likelihood ratios together to form a composite likelihood ratio.

Bayes' theorem looks deceptively basic, but turns out to encapsulate an immensely powerful way of learning from data.

Likelihood Ratios and Forensic Science

On Saturday 25 August 2012, archaeologists began an excavation for Richard III's remains by digging in a car park in Leicester. Within a few hours they found their first skeleton. What is the probability that this was Richard III?

In popular legend, promoted by the Tudor apologist William Shakespeare, Richard III (the last king of the House of York) was an evil hunchback. While this is a highly contested view, it is a matter of historical record that he was killed at the

Battle of Bosworth Field on 22 August 1485, aged 32, his death effectively ending the War of the Roses. His body was said to have been mutilated and brought for burial to Greyfriars Priory in Leicester, which was later demolished and eventually became covered by a car park.

Considering just the information provided, we might assume that this skeleton was the remains of Richard III if *all* the following were true:

- he had really been buried in Greyfriars;
- his body had not been dug up and moved or scattered in the intervening 527 years;
- the first skeleton found happened to be him.

Suppose we make rather pessimistic assumptions, and assume only a 50% probability that the stories of his burial were true, and a 50% probability that his skeleton is still where he was originally buried in Greyfriars. And imagine that up to 100 other bodies were also buried in the identified location (the archaeologists had a good idea where to dig, since Richard had been reported to have been buried in the choir of the friary). Then the probability of all the above events being true is $1/2 \times 1/2 \times 1/100 = 1/400$. This is a fairly low chance that this skeleton is Richard III; the researchers who originally carried out this analysis assumed a 'sceptical' prior probability of 1/40, and so we are being considerably more sceptical.[1]

But when the archaeologists examined the skeleton in detail they found a remarkable series of supporting forensic findings, which included radiocarbon dating of the bones (there was a 95% probability that they dated from AD 1456 to AD 1530), the fact that it was a male of around thirty, the skeleton displayed scoliosis (curvature of the spine), and evidence that the body had been mutilated after death. Genetic analysis involving known descendants of close relatives of Richard (he had no children himself) revealed shared mitochondrial DNA (through his mother). The male Y chromosome did not support a relationship, but this could easily be explained by breaks in the male line due to mistaken paternity.

The evidential value of each item of evidence can be summarized by its likelihood ratio, which in this situation is defined as

$$\text{likelihood ratio} = \frac{\text{probability of evidence, if skeleton IS Richard III}}{\text{probability of evidence, if skeleton IS NOT Richard III}}$$

Table 11–1 shows the individual likelihood ratios for each piece of evidence, revealing that none of them are individually very convincing, although the researchers were cautious and deliberately erred on the side of lower likelihood ratios that did not favour the skeleton being Richard III. But if we assume these are independent forensic findings, then we are entitled to multiply the likelihood ratios to get an overall assessment of the strength of the combined evidence, which comes to an 'extremely strong' value of 6.7 million. The verbal terms used in the table are taken from the scale shown in Table 11.2, which has been recommended for use in court.[2]

So is this evidence convincing? Remember we calculated a conservative initial probability of 1 in 400 that this skeleton was Richard III, before taking into account the detailed

Evidence	Likelihood ratio (conservative estimate)	Verbal equivalent
Radiocarbon dating AD 1456–1530	1.8	Weak support
Age and sex of skeleton	5.3	Weak support
Scoliosis	212	Moderately strong support
Post-mortem wounds	42	Moderate support
mtDNA match	478	Moderately strong support
Y chromosome not matching	0.16	Weak evidence against
Combined evidence	6.5 million	More than extremely strong support

Table 11.1
Likelihood ratios assessed for items of evidence found on skeleton found in Leicester, comparing hypotheses that the skeleton is, or is not, Richard III. The combined likelihood ratio of 6.5 million is obtained by multiplying together all the individual likelihood ratios.

Value of Likelihood ratio	Verbal Equivalent
1 – 10	Weak support for proposition
10 – 100	Moderate support
100 – 1,000	Moderately strong support
1,000 – 10,000	Strong support
10,000 – 100,000	Very strong
100,000 – 1,000,000	Extremely strong

Table 11.2
Recommended verbal interpretations of likelihood ratios when reporting forensic findings in court.

forensic findings. This corresponds to initial odds of around 1 to 400: Bayes' theorem tells us to multiply this by the likelihood ratio to give final odds, which therefore come to 6.7 million / 400 = 16,750. So, even if we are extremely cautious indeed in assessing the prior odds and the likelihood ratios, we could say that the odds are around 17,000 to 1 that the skeleton is Richard III.

The researchers' own 'sceptical' analysis led them to posterior odds of 167,000 to 1, or a 0.999994 probability that they had found Richard III. This was considered sufficient evidence to justify burying the skeleton with full honours in Leicester Cathedral.

In legal cases, likelihood ratios are typically attached to DNA evidence in which a 'match' of some degree is found between the suspect's DNA and a trace found at the scene of the crime. The two competing hypotheses are that either the suspect left the trace of DNA, or someone else did, so that we can express the likelihood ratio as follows:

$$\text{likelihood ratio} = \frac{\text{probability of DNA match,}}{\text{probability of DNA match,}}$$
$$\frac{\text{assuming suspect left the trace}}{\text{assuming someone else left the trace}}$$

The number on the top of this ratio is generally taken to be one, and the number on the bottom is assumed to be the chance that a random person picked from the population would coincidentally provide the match – this is known as the **random match probability**. Typical likelihood ratios for DNA evidence can be in the millions or billions, although the exact values might be contested, such as when there are complications due to the traces containing a mix of DNA from multiple people.

Individual likelihood ratios are allowed in British courts, but they cannot be multiplied up, as in the case of Richard III, since the process of combining separate pieces of evidence is supposed to be left to the jury.[3] The legal system is apparently not yet ready to embrace scientific logic.

Would the Archbishop of Canterbury cheat at poker?

It is a lesser known fact about the renowned economist John Maynard Keynes that he studied probability, and came up with a thought experiment to illustrate the importance of taking into account the initial odds when assessing the implications of evidence. In this exercise, he asked us to imagine playing poker with the Archbishop of Canterbury, who in the first round deals himself a winning royal flush. Should we suspect him of cheating?

The likelihood ratio for this event is

$$\text{likelihood ratio} = \frac{\text{probability of royal flush, assuming Archbishop cheating}}{\text{probability of royal flush, assuming Archbishop just lucky}}$$

We might assume the numerator is one, while the denominator can be calculated to be 1/72,000, making the likelihood ratio 72,000 – using the standards in Table 11.2, this corresponds to 'very strong' evidence that the Archbishop is cheating. But should we conclude that he is truly cheating? Bayes' theorem tells us that our final odds should be based on the product of this likelihood ratio with the initial odds. It seems reasonable to assume that, at least before we started playing, we would place strong odds against the Archbishop cheating, perhaps 1 to 1,000,000, given that he is supposed to be a respectable man of the cloth. So the product of the likelihood ratio and the prior odds ends up being around 72,000/1,000,000, which are odds of around 7/100, corresponding to a probability of 7/107 or 7% that he is a cheat. So we should give him the benefit of the doubt at this stage, whereas we might not be so generous with someone we had just met in the pub. And perhaps we should keep a careful eye on the Archbishop.

Bayesian Statistical Inference

Bayes' theorem, even if it is not permitted in UK courts, is the scientifically correct way to change our mind on the basis of new evidence. Expected frequencies make Bayesian analysis reasonably straightforward for simple situations that involve only two hypotheses, say about whether someone does or does not have a disease, or has or has not committed an offence. However, things get trickier when we want to apply the same ideas to drawing inferences about unknown quantities that might take on a range of values, such as parameters in statistical models.

The Reverend Thomas Bayes' original paper in 1763 set out to answer a very basic question of this nature: given something has happened or not happened on a known number of similar occasions, what probability should we give to it happening next time?* For example, if a thumbtack has been flipped 20 times and it has come down point-up 15 times and point-down 5 times, what is the probability of it landing point-down next time? You might by now think the answer obvious: 15/20 = 75%. But this might not be the Reverend's answer – he might say 16/22 = 73%. How would he come to this?

Bayes used a metaphor of a billiard table† which is hidden from your view. Suppose a white ball is thrown at random on to the table, its position along the table marked with a line, and then the white ball is removed. A number of red balls are then thrown at random on to the table, and you are told only how many lie to the left and how many to the right of the line. Where do you think the line might be, and what should be your probability of the next red ball falling to the left of the line?

For example, say five red balls are thrown, and we are told that two landed to the left, and three to the right of the line left by the white ball, as shown in Figure 11.4(a). Bayes showed that our beliefs about the position of the line should be described by the probability distribution shown in Figure 11.4(b) – the mathematics is quite complex and is given in an endnote.[4] The position of the dashed line, which

---

* His exact words were, 'Given the number of times on which an unknown event has happened and failed: Required the chance that the probability of its happening in a single trial lies somewhere between any two degrees of probability that can be named', which is reasonably clear, except in modern terminology we would probably reverse his use of 'chance' and 'probability'.

† Being a Presbyterian minister, he just called it a table.

(a)

(b)

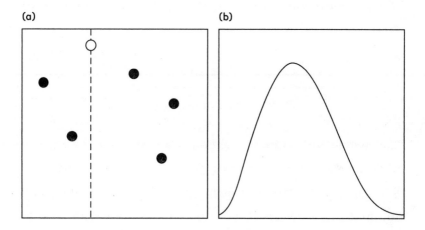

Figure 11.4
Bayes' 'billiard' table. (a) A white ball is thrown on to the table and the dashed line indicates its resting position. Five red balls are thrown on to the table and land as shown. (b) An observer cannot see the table, and is only told that two red balls landed to the left, and three on the right of the dashed line. The curve represents the observer's probability distribution for where the white ball landed, superimposed on the billiard table. The curve's mean is 3/7, which is also the observer's current probability for the next red ball to land on the left of the line.

indicates where the white ball landed, is estimated to be 3/7 along the table, which is the mean (expectation) of this distribution.

This value of 3/7 may seem odd, as the intuitive estimate might be 2/5 – the proportion of red balls landing to the left of the line. Instead Bayes showed that in these circumstances we should estimate the position as

$$\frac{\text{Number of red balls lying to the left} + 1}{\text{Total number of red balls} + 2}$$

This means, for example, that before any red balls are thrown at all, we can estimate the position to be $(0 + 1)/(0 + 2) = \frac{1}{2}$, whereas the intuitive approach might suggest that we could not give any answer since there is not yet any data. Essentially Bayes is making use of the information about how the position of the line has been initially decided, since we know it is picked at random by throwing the white ball. This initial information takes the same role as the prevalence used in breast screening or dope-testing – it is known as prior information and it influences our final conclusions. In fact, since Bayes' formula adds one to the number of red balls to the left of the line and two to the total number of red balls, we might think of it as being equivalent to having already thrown two 'imaginary' red balls, and one having landed each side of the dashed line.

Note that if none of the five balls had landed to the left of the dashed line, we would not have estimated its position as 0/5, but instead as 1/7, which seems much more sensible. The Bayes' estimate can never be 0 or 1, and is always nearer to ½ than the simple proportion: this is known as **shrinkage**, in that estimates are always pulled in or shrunk, towards the centre of the initial distribution, in this case ½.

Bayesian analysis uses knowledge about how the position of the dashed line was decided to establish a **prior distribution** for its position, combines it with evidence from the data known as the **likelihood,** to give a final conclusion known as the **posterior distribution,** which expresses all we currently believe about the unknown quantity. So, for example, computer software can calculate that an interval that runs from 0.12 to 0.78 contains 95% of the probability in Figure 11.4(b), and so we can claim with 95% certainty that the line marking the white ball is between these limits. This interval will become steadily narrower as more and more red balls are thrown on to the table and their positions relative to the line announced, until eventually we shall converge on the correct answer.

The main controversy about Bayesian analysis is the source of the prior distribution. In Bayes' billiard table, the white ball was thrown at random on to the table and so everyone would agree that the prior distribution is uniformly spread over the whole line between 0 and 1. When this kind of physical knowledge is unavailable, suggestions for obtaining prior distributions include using subjective judgement, learning from historical data, and specifying **objective priors** that try to let the data speak for themselves without introducing subjective judgement.

Perhaps the most important insight is that there is no 'true' prior distribution, and any analysis should include a sensitivity analysis to a number of alternative choices, encompassing a range of different possible opinions.

How can we better analyse pre-election polls?

We have seen how Bayesian analysis provides a formal mechanism for using background knowledge to make more realistic inferences about the particular problem in front of us. These ideas can (literally) be taken to another level, in that multi-level or **hierarchical modelling** simultaneously analyses various individual quantities: the power of these models can be seen in successes in pre-election polls.

We know that polls should ideally be based on large, random, representative samples, but these are increasingly expensive to establish, and people are in any case increasingly unwilling to respond to survey questions. So polling companies are now generally relying on online panels – these are known not to be truly representative, but sophisticated statistical modelling is then used to try to work out what might have been the responses if the companies had been able to take a proper random sample. The traditional warning against trying to make a silk purse out of a sow's ear might come to mind.

Things get even worse when it comes to pre-election polls, since attitudes will not be uniform across a country and so claims about the overall national picture should come from aggregating results from many different states or constituencies. We ideally need to make conclusions at the local level, but the people on the online panel will be sparsely scattered in a non-random fashion across these local areas, meaning that there is very limited data on which to base local analyses.

The Bayesian response to this problem is known as **multi-level regression and post-stratification (MRP)**. The basic idea is to break down all possible voters into small 'cells', each comprising a highly homogeneous group of people – say living in the same area, with the same age, gender, past voting behaviour, and other measurable characteristics. We can use background demographic data to estimate the number of people in each of these cells, and these are all assumed to have the same probability of voting for a certain party. The problem is working out what this probability is, when our non-random data may mean that we only have a few people, or maybe none at all, in a particular cell.

The first step is to build a regression model for the probability of voting a particular way given the characteristics of the cell, so our problem is reduced to estimating the coefficients of the regression equation. But there are still too many coefficients to estimate reliably using the standard methods, and this is where Bayesian ideas come in. Coefficients corresponding to different areas are assumed to be *similar*, a sort of intermediate point between assuming they are precisely the same, and assuming they are utterly unrelated.

Mathematically, this assumption can be shown to be equivalent to assuming all these unknown quantities have been drawn from the same prior distribution, and this enables us to move many individual, rather imprecise, estimates towards each other, which results in smoother, more confident conclusions that are not so influenced by a few odd observations. Having made these more robust estimates of the voting behaviour within each of the thousands of cells, the results can be combined to produce a prediction for how the whole country will vote.

In the 2016 US Presidential election, polls based on multi-level regression and post-stratification gave the correct winner in 50 of 51 states and the District of Columbia, getting only Michigan wrong, on the basis of interviews of only 9,485 voters in the weeks preceding the election. Similar good predictions came in the 2017 UK election, where the polling company YouGov interviewed 50,000 people a week without being concerned with obtaining a representative sample, but then used MRP to predict a hung parliament with the Conservatives obtaining 42% of the vote, which is exactly what happened. Polls using more traditional methods failed spectacularly.[5]

So can we make the proverbial silk purse out of a convenient non-random sow's ear? MRP is no panacea – if a large number of respondents give systematically misleading answers and so do not represent their 'cell', then no amount of sophisticated statistical analyses will counter that bias. But it appears to be beneficial to use Bayesian modelling of every single voting area, and we shall see later that this has been spectacularly successful in exit polls conducted on the day of elections.

Bayesian 'smoothing' can bring precision to very sparse data, and the techniques are being increasingly used for modelling, for example, how diseases spread over space and time.

Bayesian learning is also now seen as a fundamental process of human awareness of the environment, in that we have prior expectations about what we will see in any context, and then only need to take notice of unexpected features in our vision which are then used to update our current perceptions. This is the idea behind the so-called Bayesian Brain.[6] The same learning procedures have been implemented in self-driving cars, which have a probabilistic 'mental map' of their surroundings that is constantly being updated by recognition of traffic lights, people, other cars, and so on: 'In essence, a robot car "thinks" of itself as a blob of probability, traveling down a Bayesian road.'[7]

These problems are about estimating quantities that describe the world, but using Bayesian methods for assessing scientific hypotheses remains more controversial. Just as in Neyman–Pearson testing, we first need to set up two competing hypotheses. A null hypothesis H_0, which is usually the absence of something, such as there being no Higgs boson, or a medical treatment having no effect. The alternative hypothesis, H_1, says that something important exists.

The ideas behind Bayesian hypothesis testing are then essentially the same as for legal cases, in which the null hypothesis is generally innocence, the alternative is guilt, and we express the relative support that a piece of evidence provides for these two hypotheses by the likelihood ratio. For scientific hypothesis testing, the precise equivalent to the likelihood ratio is the **Bayes factor**, with the difference that scientific hypotheses generally contain unknown parameters, such as the true effect under the alternative hypothesis. The

Bayes factor can only be obtained by averaging with respect to the prior distribution of the unknown parameters, which makes the prior distribution – the most controversial part of a Bayesian analysis – crucially important. So attempts to replace standard significance testing with Bayes factors, particularly in psychology, are a source of considerable argument, with critics pointing out that behind any Bayes factor are lurking assumed prior distributions for any unknown parameters in both the null and alternative hypotheses.

Robert Kass and Adrian Raftery are two renowned Bayesian statisticians who proposed a widely used scale for Bayes factors, shown in Table 11.3. Note the contrast to the scale in Table 11.2 for verbally interpreting likelihood ratios for legal cases, where a likelihood ratio of 10,000 was required to declare the evidence as 'very strong', in contrast to scientific hypotheses only needing a Bayes factor of greater than 150. This perhaps reflects the need to establish criminal guilt 'beyond reasonable doubt', whereas scientific claims are made on weaker evidence, with many being overturned on further research.

Our chapter on hypothesis testing contained a claim that a P-value of 0.05 was only equivalent to 'weak evidence'. The reasoning for this is partly based on Bayes factors: $P = 0.05$ can be shown to correspond, under some reasonable priors under the alternative hypothesis, to Bayes factors between 2.4 and 3.4, which Table 11.3 suggests is weak evidence. As we saw in Chapter 10, this led to a proposed reduction in the necessary P-value for claiming a 'discovery' to 0.005.

Unlike null-hypothesis significance testing, Bayes factors treat the two hypotheses symmetrically, and so can actively

Bayes factor	Strength of evidence
1 to 3	not worth more than a bare mention
3 to 20	positive
20 to 150	strong
>150	very strong

Table 11.3
Kass and Raftery's scale for interpretation of Bayes factors in favour of a hypothesis.[8]

support a null hypothesis. And if we are willing to put prior probabilities on hypotheses, we might even calculate posterior probabilities of alternative theories for how the world works. Suppose, based on theoretical grounds alone, we judged it 50:50 whether the Higgs boson existed, corresponding to prior odds of 1. The data discussed in the last chapter gave a P-value of around 1/3,500,000, and this can be converted to a maximum Bayes factor of around 80,000 in favour of the Higgs boson, which is very strong evidence even according to legal usage.

When combined with prior odds of 1, this turns into posterior odds of 80,000 to 1 for the existence of the Higgs boson, or a probability of 0.99999. But neither the legal nor scientific community generally approve of this kind of analysis, even if it can be used for Richard III.

An Ideological Battle

In this book we have moved from the informal examination of data, through communication with summary statistics, to using probability models to arrive at confidence intervals, P-values and so on. These standard inferential tools, with which generations of students have occasionally struggled, are known as 'classical' or 'frequentist' methods, since they are based on long-run sampling properties of statistics.

The alternative Bayesian approach is based on fundamentally different principles. As we have seen, external evidence about the unknown quantities, expressed as a prior distribution, is combined with evidence from the underlying probability model for the data, known as the likelihood, to give a final, posterior distribution which forms the basis for all conclusions.

If we seriously adopt this statistical philosophy, the sampling properties of statistics become irrelevant. And so having spent years learning that a 95% confidence interval does not mean there is a 95% probability that the true value lies in the interval,* the poor student now has to forget all that: a Bayesian 95% uncertainty interval has precisely the latter meaning.

But the argument about the 'correct' way to do statistical inference is even more complex than a simple dispute between frequentists and Bayesians. Just like political movements, each school splits into multiple factions who have often been in conflict with each other.

In the 1930s, a three-cornered fight erupted into the public arena. The forum was the Royal Statistical Society, which then as now meticulously recorded and published the discussion of papers presented at its meetings. When Jerzy Neyman proposed his theory of confidence intervals in 1934, Arthur Bowley, a strong advocate of the method of the Bayesian approach, then known as inverse probability, said, 'I am not at all sure that the "confidence" is not a "confidence trick",' and followed this by suggesting a Bayesian approach was necessary: 'Does that really take us any further? . . . Does it really lead us towards what we need – the chance that in the universe which we are sampling the proportion

---

* Remember, it means that, in the long run, 95% of such intervals will contain the true value – but we can't say anything about any particular interval.

is within ... certain limits? I think it does not.' The derisive linking of confidence intervals with confidence tricks continued in the subsequent decades.

The following year, in 1935, open warfare then broke out between two non-Bayesian camps, with Ronald Fisher on one side, and Jerzy Neyman and Egon Pearson on the other. The Fisherian approach was based on estimation using the 'likelihood' function, which expresses the relative support given to the different parameter values by the data, and hypothesis testing was based on P-values. In contrast, the Neyman–Pearson approach, which as we have seen was known as 'inductive behaviour', was very much focused on decision-making: if you decide the true answer is in a 95% confidence interval, then you will be right 95% of the time, and you should control Type I and Type II errors when hypothesis testing. They even suggested you should 'accept' the null hypothesis when it was included in the 95% confidence interval, a concept that was anathema to Fisher (and has subsequently been rejected by the statistical community).

Fisher first accused Neyman 'of falling into the series of misunderstandings which his paper revealed'. Pearson then rose to Neyman's defence, saying that 'while he knew there was a widespread belief in Professor Fisher's infallibility, he must, in the first place, beg leave to question the wisdom of accusing a fellow-worker of incompetence without, at the same time, showing that he had succeeded in mastering the argument.' The acrimonious dispute between Fisher and Neyman continued for decades.

The struggle for statistical ideological supremacy continued after the Second World War, but over time the more standard, non-Bayesian schools have resolved into a pragmatic mix, with experiments generally designed using a Neyman–Pearson approach of Type I and Type II errors, but then analysed from a Fisherian perspective using P-values as measures of evidence. As we have seen in the context of clinical trials, this strange amalgam seems to work fairly well, leading prominent (Bayesian) statistician Jerome Cornfield to remark, 'the paradox is that a solid structure of permanent value has, nevertheless, emerged, lacking only the firm logical foundation on which it was originally thought to have been built.'[9]

The purported advantages of conventional statistical methods over Bayesianism include the apparent separation of the evidence in the data from subjective factors; general ease in computation; wide acceptability and established criteria for 'significance'; availability of software; and existence of robust methods that do not have to make strong assumptions about the shape of distributions. Whereas Bayesian enthusiasts would claim that the very ability to make use of external, and even explicitly subjective, elements is what enables more powerful inferences and predictions to be made.

The statistical community used to engage in lengthy vituperative arguments about the foundations of the subject, but now a guarded truce has been called and a more ecumenical approach is the norm, with methods chosen according to the practical context rather than their ideological credentials derived from Fisher, Neyman–Pearson or Bayes. This seems a sensible and pragmatic compromise in an argument that can appear somewhat obscure to non-statisticians. My personal view is that, while they may well disagree about the fundamentals of their subject, reasonable statisticians will generally come to similar conclusions. The problems that arise in statistical science do not generally come from the philosophy underlying the precise methods that are used. Instead, they are more likely to be due to inadequate design, biased data, inappropriate assumptions and, perhaps most important, poor scientific practice. And in the next chapter we shall take a look at this dark side of statistics.*

---

*   But I still prefer the Bayesian approach.

Summary

- Bayesian methods combine evidence from data (summarized by the likelihood) with initial beliefs (known as the prior distribution) to produce a posterior probability distribution for the unknown quantity.
- Bayes' theorem for two competing hypotheses can be expressed as posterior odds = likelihood ratio $\times$ prior odds.
- The likelihood ratio expresses the relative support for two hypotheses from an item of evidence, and is sometimes used to summarize forensic evidence in criminal trials.
- When the prior distribution comes from some physical sampling process, Bayesian methods are uncontroversial. However generally a degree of judgement is necessary.
- Hierarchical models allow evidence to be pooled across multiple small analyses that are assumed to have parameters in common.
- Bayes factors are the equivalent of likelihood ratios for scientific hypotheses, and are a controversial substitute for null-hypothesis significance testing.
- The theory of statistical inference has a long history of controversy, but issues of quality of data and scientific reliability are more important.

CHAPTER 12

How Things Go Wrong

Does extra-sensory perception (ESP) exist?

In 2011, the eminent American social psychologist Daryl Bem published a major paper in a prominent psychology journal that featured the following experiment. A hundred students sat in front of a computer screen showing two curtains, and chose which of either the left or right curtain hid an image. The curtains then 'opened' to reveal whether they were correct or not, and this was repeated for a series of 36 images. The twist was that, unknown to the participants, the position of the image was determined at random *after* the subject had made their choice, and so any excess correct choices over chance would be ascribed to *precognition* of where the image would appear.

Bem reported that, instead of the expected 50% success rate under the null hypothesis of no precognition, subjects chose correctly 53% of the time when an erotic image was shown ($P = 0.01$). The paper contained the results of eight further experiments in precognition, with over 1,000 participants and spread over ten years, and he observed statistically significant results in favour of precognition in eight of the nine studies. Is this convincing proof that extra-sensory perception (ESP) exists?

This book has, I hope, illustrated some powerful applications of statistical science in solving real-world problems, carried out with skill and care by practitioners who are mindful of its limitations and potential pitfalls. But the real world is not always so worthy of admiration. It is now time to look at what happens when the science and art of statistics are not carried out so well. We shall then look at how Bem's paper was received and critiqued.

There is a reason so much attention is now being paid to poor-quality statistical practice: it has been blamed for what is known as the **reproducibility crisis** in science.

The 'Reproducibility Crisis'

Chapter 10 explored John Ioannidis's notorious 2005 claim that most published research findings were false, and since then many other researchers have argued that there is a fundamental lack of reliability in the published scientific literature. Scientists have failed to replicate studies done by their peers, suggesting that the original studies are not as trustworthy as previously thought. Although initially focused on medicine and biology, these accusations have since spread to psychology and other social sciences, although the actual percentage of claims that are either exaggerated or false is contested.

Ioannadis's original claim was based on a theoretical model, but an alternative approach is to take past studies and try to replicate them, in the sense of conducting similar experiments and seeing if similar results are observed. The Reproducibility Project was a major collaboration in which 100 psychological studies were repeated with larger sample sizes, and so had higher power to detect a true effect if it existed. The project revealed that whereas 97% of the original studies had statistically significant results, only 36% of the replications did.[1]

Unfortunately, this was widely reported as implying that the remaining 63% of 'significant' studies were false claims, but this falls into the trap of making a strict division between studies that are either significant or not-significant. Distinguished US statistician and blogger Andrew Gelman has pointed out that 'the difference between "significant" and "not significant" is not itself statistically significant'.[2] In fact only 23% of original and replication studies had results that were significantly different from each other, which is perhaps a more appropriate estimate of the proportion of the original studies with exaggerated or false claims.

Rather than thinking in terms of significant or not-significant as determining a 'discovery', it would be better to focus on the sizes of the estimated effects. The Reproducibility Project found that replication effects were on average in the same direction as the original studies, but were around half their magnitude. This points to an important bias in the scientific literature: a study which has found something 'big', at least some of which is likely to have been luck, is likely to lead to a prominent publication. In an analogy to regression to the mean, this might be termed 'regression to the null', where early exaggerated estimates of effects later decrease in magnitude towards the null hypothesis.

The claimed reproducibility crisis is a complex issue, rooted in the excessive pressure put on researchers to make 'discoveries' and publish their results in prestigious scientific journals, all of which is crucially dependent on finding statistically significant results. No single institution or profession is to blame. We have also showed when discussing hypothesis testing that, even if statistical practice were perfect, the rarity of true and substantial effects means a substantial proportion of results that are claimed to be 'significant' are inevitably going to be false-positives (Figure 10.5). But, as we now see, statistical practice is often far from perfect.

Statistics can be done badly at every stage of the PPDAC cycle. Right from the beginning, we may try to tackle a Problem that just cannot be answered with the information available. For example, if we set out to find out why teenage pregnancy rates have fallen so dramatically in the UK over the last decade, nothing in the observed data can offer an explanation.*

Then the Planning can go wrong, for example by

- Choosing a sample that is convenient and inexpensive rather than representative, for example telephone polls before elections.
- Asking leading questions or using misleading wording in surveys, such as 'How much do you think you can save by buying online?'
- Failing to make a fair comparison, such as assessing homeopathy by only observing volunteers for the therapy.
- Designing a study that is too small and so has low power, which means that fewer true alternative hypotheses are detected.

---

* The fall began soon after the start of Facebook, but the data cannot tell us whether this is correlation or causation.

- Failing to collect data on potential confounders, lack of blinding in randomized trials, and so on.

As Ronald Fisher famously put it, 'To consult the statistician after an experiment is finished is often merely to ask him to conduct a post mortem examination. He can perhaps say what the experiment died of.'[3]

When it comes to collecting Data, common problems include excessive missing responses, people dropping out of the study, recruitment being much slower than anticipated, and simply getting everything coded up efficiently. All these issues should have been foreseen and avoided by careful piloting.

The easiest way for Analysis to go wrong is simply to make a mistake. Many of us will have made errors in coding or spreadsheets, but perhaps not with the consequences of the following examples:

- Prominent economists Carmen Reinhart and Kenneth Rogoff published a paper in 2010 which strongly influenced attitudes to austerity. A PhD student later found that five countries had been inadvertently left out of their main analysis due to a simple spreadsheet error.*[4]
- A programmer for AXA Rosenberg, a global equity investment firm, incorrectly programmed a statistical model so that some of its calculated risk elements were too small by a factor of ten thousand, leading to

---

* This error, in combination with other criticisms, was claimed to change conclusions of the study, but this is strongly disputed by the original authors.

$217 million in losses to clients. In 2011 the Securities
and Exchange Commission (SEC) fined AXA Rosenberg
for this amount plus an additional $25 million in
penalties, resulting in a $242 million fine for not
reporting to investors a risk model error.[5]

Calculations might be computationally correct, but use incorrect statistical methods. Some popular contenders for inappropriate methods include

- Carrying out a 'cluster randomized' trial in which
  whole groups of people, such as all patients in a general
  practice, have been simultaneously randomly allocated
  to a particular intervention, and yet analysing the data
  as if the people had been individually randomized.

- Measuring two groups at baseline and after an intervention,
  and saying the groups are different if one is significantly
  changed from their baseline, and the other group's change
  is not significant. The correct procedure is to carry out a
  formal statistical test of whether the groups differ – this is
  known as a test of interaction.

- Interpreting 'non-significant' as meaning 'no effect'. For
  example, in the alcohol and mortality study mentioned
  in Chapter 10, males aged 50–64 and drinking 15–20 units
  of alcohol a week had a significantly reduced mortality
  risk, whereas the reduction in men drinking a little less
  or a little more was not significantly different from
  zero. This was claimed as an important difference in the
  paper, but the confidence intervals revealed there was
  negligible difference between these groups. Once again,

the difference between significant and not-significant is
not necessarily significant.

When it comes to drawing Conclusions, perhaps the most
blatant form of poor practice is to carry out many statistical tests, and then only report the most significant ones and
interpret these at face value. We have seen that this hugely
increases the chance of finding a significant P-value, even
bringing life to a dead fish. It is the equivalent of only tele-
vising the goals a team scores, and not the goals they let in:
it is impossible to get a true impression when there is such
selective reporting.

Selective reporting starts to cross the boundary between
simple incompetence and scientific misconduct, and there
is disturbing evidence that this is not uncommon. In the
US, there has even been a criminal conviction for selec-
tively reporting significant results in a subset analysis. Scott
Harkonen was the CEO of InterMune, a company which
carried out a clinical trial of their new drug for idiopath-
ic pulmonary fibrosis. The trial showed no overall benefit,
but a significantly reduced mortality in the small subset of
patients with mild-to-moderate disease. Harkonen issued a
press release to investors reporting this result, indicating he
believed the study would lead to huge sales. While he didn't
say anything demonstrably untrue, in 2009 a jury convicted
him of wire fraud, with specific intent to defraud investors.
The government had sought a 10-year prison sentence and
a $20,000 fine, but he was sentenced to six months' house
arrest and three years' probation. A subsequent clinical trial
found no benefit of the drug in this subset of patients.[6]

Statistical misconduct may or may not be a conscious decision. It has even been deliberately used to show up the inadequacies of the scientific peer-review and publishing process. Johannes Bohannon of the German Institute of Diet and Health conducted a study in which people were randomized into three groups and given either a standard, low-carb or low-carb with extra chocolate diet. They had a battery of measurements taken over three weeks, and the study concluded that the weight loss in the chocolate group exceeded that of the low-carb group by 10% (P = 0.04). This 'significant' result was submitted to a journal that deemed it an 'outstanding manuscript', and suggested that for €600 it 'could be accepted directly in our premier journal'. On publication the press release from the Institute of Diet and Health led to stories in numerous news outlets, with headlines such as 'Chocolate Accelerates Weight Loss'.

But this was then all revealed to be a deliberate fraud. 'Johannes Bohannon' was really John Bohannon, a journalist; the Institute of Diet and Health did not exist; and the only real element was the data, which was not fabricated. But there were only five subjects per group, a large number of tests were made, and only the significant differences reported.

The authors of this spurious article immediately owned up to their deception, but not all statistical frauds are done in order to show up weaknesses in the peer-review process.

Deliberate Fraud

Deliberate fabrication of data does occur, but is thought to be relatively rare. A review of anonymous self-reports estimated that 2% of scientists admitted falsification of data, while the US National Science Foundation and Office of Research Integrity deal with a fairly small number of deliberately dishonest acts, although those detected must be an underestimate.[7]

It seems entirely appropriate that statistical fraud can be detected using statistical science. Uri Simonsohn, a psychologist at the University of Pennsylvania, has examined statistics describing supposedly randomized trials that should show typical random variation, but are either implausibly similar or different. For example, he noticed that three estimated standard deviations quoted in a report, which were supposed to come from different groups of 15 individuals, were all equal to 25.11. Simonsohn obtained the raw data and showed by simulation that the chance of getting such similar standard deviations was vanishingly small – the researcher who was responsible for the report later resigned.[8]

Cyril Burt, a British psychologist who was renowned for his research on the heritability of IQ, was posthumously accused of fraud when it was found that the correlation coefficients he quoted for the IQ of twins who had been reared apart hardly changed over time in spite of a steadily increasing group of twins: the correlation was 0.770 in 1943, 0.771 in 1955, and 0.771 in 1966. He was accused of making up data, but all his records had been burned after he died. The issue is still disputed, since some claim that this must be a mistake since it is too obvious a fraud for him to commit.

It would be easier if sheer incompetence and dishonesty were the only problems with statistics, grave though they are. We could educate, check, replicate, open up data for examination, and so on, as we shall see in the final chapter on doing stats well. But there is a bigger, subtler problem, which some claim is a major contributor to the reproducibility crisis.

'Questionable Research Practices'

Even if the data has not been made up, the final analysis is appropriate, and a statistic and its accompanying P-value are numerically correct, it can be difficult to know how to interpret the results if we don't know exactly what the researchers have done in the process of arriving at their conclusions.

We have seen the problems that result when researchers only report significant findings, but perhaps more important are the conscious or unconscious set of minor decisions that might be made by the researcher depending on what the data seem to be showing. These 'tweaks' might include decisions about changes in the design of the experiment, when to stop collecting data, what data to exclude, what factors to adjust for, what groups to emphasize, what outcome measures to focus on, how to split continuous variables into groups, how to handle missing data, and so on. Simonsohn calls these decisions 'researcher degrees of freedom', while Andrew Gelman refers more poetically to the 'garden of forking paths'. All these tweaks are likely to increase the chance of getting statistical significance, and all come under the general banner of 'questionable research practices', sometimes abbreviated to QRPs.

It is important to distinguish what are known as **exploratory** and **confirmatory studies**. Exploratory experiments are just what they say: flexible investigations intended to look at many possibilities and suggest hypotheses to test later in more formal, confirmatory studies. Any amount of tweaking is fine in exploratory studies, but confirmatory studies should be carried out according to a pre-specified, and preferably public, protocol. Each can use P-values to summarize the strength of evidence for their conclusions, but these P-values should be clearly distinguished and interpreted very differently.

Activities that are intended to create statistically significant results have come to be known as 'P-hacking', and although the most obvious technique is to carry out multiple tests and report the most significant, there are many more subtle ways in which researchers can exercise their degrees of freedom.

> Does listening to the Beatles' song 'When I'm Sixty-Four' make you younger?

You might feel fairly confident about the correct answer to this question. Which makes it all the more impressive that Simonsohn and colleagues managed, admittedly by some fairly devious means, to get a significant positive result.[9]

University of Pennsylvania undergraduates were randomized to listen to either 'When I'm Sixty-Four' by the Beatles, or 'Kalimba', or 'Hot Potato' by the Wiggles. Then the students were asked when they were born, how old they felt, and a series of delightfully irrelevant questions.*

---

* These included how much they would enjoy eating at a diner, the square root of 100, their agreement with the statement 'computers are complicated machines', their father's age, their mother's age, whether they would take advantage of an early-bird special, their political orientation, which of four Canadian quarterbacks they believed won an award, how often they refer to the past as 'the good old days', and so on.

Simonsohn and his colleagues repeatedly analysed the data in every way they could think of, and kept enrolling participants until they found some kind of significant association. This happened after 34 subjects, and although there was no significant relationship between the age of the participants and the record they listened to, by only comparing 'When I'm Sixty-Four' and 'Kalimba', they managed to get P < 0.05 in a regression that adjusted for father's age. Naturally, they only reported the significant analysis without at first mentioning the vast number of tweaks, fiddles and selective reporting that had gone on – these were revealed at the end of the paper, which has become a classic deliberate demonstration of the practice of what is now known as HARKing – inventing the Hypotheses After the Results are Known.

How Much Do People Actually Engage in These Questionable Research Practices?

In a 2012 survey of 2,155 US academic psychologists,[10] just 2% admitted to falsifying data. But when asked about a list of ten questionable research practices,

- 35% said that they had reported an unexpected finding as having been predicted from the start.
- 58% said they had carried on collecting more data after seeing whether the results were significant.
- 67% said they had failed to report all of a study's responses.
- 94% acknowledged at least one of the questionable research practices that had been listed.

They generally argued that these practices were defensible – after all, why not report an interesting, although unexpected, finding? Again, the problems arise due to a blurring of boundaries between exploratory and confirmatory studies: many of the practices, including HARKing, may be fine in an exploratory study that is deliberately intended to develop new ideas to test, but should be strictly forbidden in studies that claim to prove anything.

Communication Breakdown

Whether the statistical work is good or not so good, at some point it must be communicated to audiences, whether fellow professionals or a more general public. Scientists are not the only people reporting claims based on statistical evidence. Governments, politicians, charities and other non-governmental organizations are all competing for our attention, using numbers and science to provide an apparently 'objective' basis for their assertions. Technology has changed, encouraging an increasing diversity of sources to use online and social media to communicate, with few controls to ensure the reliable use of evidence.

Figure 12.1 provides a highly simplified view of the process by which we hear about statistical evidence.[11] The pipeline starts with the originators of the data, and then goes through the 'authorities', then through their press and communication offices, to the journalists who write the stories and the editors who add the headlines, and finally to us as individual members of society. Errors and distortions can occur through the whole process.

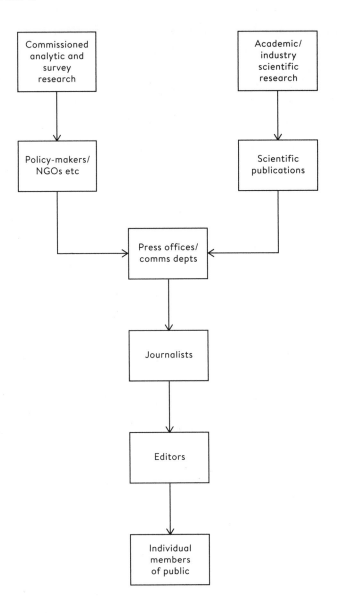

Figure 12.1
A simplistic diagram of the traditional information flows from statistical sources through to the public. At each stage there are filters arising from questionable research, interpretation and communication practices, such as selective reporting, lack of context, exaggeration of importance, and so on.

What Appears in the Literature?

The first filter occurs at the publication of the statistical work carried out by researchers. Many studies are not submitted for publication at all, either because the findings did not appear 'interesting' enough, or because they did not fit with the aims of the research organization: pharmaceutical companies, in particular, have been accused in the past of hiding studies whose outcomes did not suit them. This leaves valuable data sitting in the 'file drawer', and creates a positive bias to what appears in the literature. We do not know what we are not being told.

This positive bias is made worse by 'discoveries' that are more likely to be accepted for publication in more prominent journals, an unwillingness to publish replications, and of course all the questionable research practices that we have seen can lead to exaggerated statistical significance.

The Press Office

More potential problems arise at the next stage in the pipeline when scientific stories are passed to press offices to try to obtain media coverage. We have already seen how an over-enthusiastic press release for a study on socio-economic position and the risk of brain tumours led to the classic headline 'Why Going to University Increases Risk of Getting a Brain Tumour'. That press office is not alone in its exaggeration: a study found that of 462 press releases from UK universities in 2011:

- 40% contained exaggerated advice
- 33% contained exaggerated causal claims

- 36% contained exaggerated inference to humans from animal research
- the majority of exaggerations appearing in the press could be traced back to the press release

The same team found slightly more reassuring results in 534 press releases from major biomedical journals: causal claims or advice in a paper were exaggerated in 21% of corresponding press releases, although these exaggerations, which tended to be reported, did not produce more press coverage.[12]

We saw in Chapter 1 that the 'framing' of numbers can influence their interpretation: for example '90% fat-free' sounds somewhat better than '10% fat'. A fine example of imaginative storytelling happened when a worthy but rather dull study found that 10% of people carried a gene which protected them against high blood pressure. The communications team reframed this as 'nine in ten people carry a gene which increases the risk of high blood pressure': this negatively-framed message duly received international press coverage.[13]

The Media

Journalists tend to get the blame for poor coverage of scientific and statistical stories, but they are to a great extent at the mercy of what is fed to them in press releases and scientific papers, and how the editor's headline subsequently frames their story: few newspaper readers realize that the person who wrote the article generally has minimal control over the headline, and headlines are of course there to attract readers.

The main problem in media coverage is not outright untruths, but manipulation and exaggeration through inappropriate interpretation of 'facts': these may be technically correct, but are distorted by what we might call 'questionable interpretation and communication practices'. Here is a short list of ways the media can spice up their coverage of statistical stories. Many of these questionable practices would be seen as defensible by those whose professional career depends on attracting readers, listeners or clicks.

1. Pick stories that go against current consensus.
2. Promote stories regardless of research quality.
3. Fail to report uncertainties.
4. Fail to provide context or comparative perspective, such as a long term trend.
5. Suggest a cause when only an association is observed.
6. Exaggerate the relevance and importance of findings.
7. Claim the evidence supports a particular policy.
8. Use positive or negative framing depending on whether the aim is to reassure or frighten.
9. Neglect conflicts of interest or alternative views.
10. Use a vivid but uninformative graphic.
11. Only provide relative and not absolute risks.

The final practice is almost universal. We saw in Chapter 1 how a story about bacon increasing the risk of bowel cancer could be made to sound impressive by quoting relative rather than absolute risks. Journalists know that relative risks, often referred to by the media as simply an 'increased risk' regardless of magnitude, are an effective way of making a story look more exciting, and this is not helped by the fact that relative risks in the form of odds ratios, **rate ratios** and **hazard ratios** are the standard output from most biomedical studies.

The gripping headline 'Why Binge Watching Your TV Box-Sets Could Kill You' arose from an epidemiological study that estimated an adjusted relative risk of 2.5 for a fatal pulmonary embolism associated with watching more than 5 hours TV a night compared with less than 2.5 hours. But careful scrutiny of the absolute rate in the high-risk group (13 in 158,000 person-years) could be translated as meaning you can expect to watch more than 5 hours TV a night for 12,000 years before experiencing the event. This somewhat lessens the impact.[14]

This headline was written to attract attention and clicks, and duly succeeded – I certainly found it irresistible. When we all seek novelty and immediate stimulation, it is not surprising that the media spice up stories of studies and favour the unusual (and probably exaggerated) claim over solid statistical evidence.* In the next chapter we will see how things might be improved, but first we return to Daryl Bem's remarkable claims about precognition.

Daryl Bem knew he was publishing extraordinary claims, and to his great credit he actively encouraged replication and provided the materials to do so. However, when other researchers took him up on his challenge, and tried (and failed) to reproduce his results, the journal that had published Bem's original study refused to publish the failed replications.

So how did Bem arrive at his results? There were numerous

---

* I sometimes follow what could be called the 'Groucho principle', after Groucho Marx's paradoxical claim that he would never join a club that would have him as a member. Because stories have gone through so many filters that encourage distortion and selection, the very fact that I am hearing a claim based on statistics is reason to disbelieve it.

points at which he adjusted the designs in response to the data, and chose to highlight particular groups – for example, reporting the positive precognition when showing erotic pictures and not the negative results from non-erotic pictures. Bem has acknowledged, 'I would start one [experiment], and if it just wasn't going anywhere, I would abandon it and restart it with changes'. Some of these changes were reported in the article; others were not.*[15] Andrew Gelman observed that Bem's

> conclusions are based on P-values, which are statements regarding what the data summaries would look like, had the data come out differently, but Bem offers no evidence that, had the data come out differently, his analyses would've been the same. Indeed, the nine studies of his paper feature all sorts of different data analyses.†

His case is a classic example of someone exploiting too many researcher degrees-of-freedom. But Bem did a great service for psychology and science in general: his 2011 paper was a catalyst for collective soul-searching among scientists about the possible reasons for the lack of reliability of the scientific literature. It has even been suggested that the whole exercise, like other studies featured in this chapter, was deliberately planned by Bem to reveal the weaknesses in psychological research.

---

* Bem is quoted in an online article as saying, 'I'm all for rigor . . . but I prefer other people do it. I see its importance – have fun for some people – but I don't have the patience for it . . . If you looked at all my past experiments, they were always rhetorical devices. I gathered data to show how my point would be made. I used data as a point of persuasion, and I never really worried about, "Will this replicate or will this not?" '

† Gelman's pithy summary was that 'Bem's study was crap.'

Summary

- Poor statistical practice has some responsibility for the crisis in the reproducibility of science.
- Deliberate fabrication of data appears to be fairly rare, but errors in statistical methods are frequent.
- An even greater problem is questionable research practices that tend to lead to exaggerated claims of statistical significance.
- In the pipeline by which statistical evidence reaches the public, press offices, journalists and editors add to the flow of unjustified statistical claims through their use of questionable interpretation and communication practices.

How We Can Do Statistics Better

> What is the benefit of screening for ovarian cancer?

In 2015 a huge UK trial of ovarian cancer screening published its results. It had started back in 2001 when, after a careful power calculation, over 200,000 women were randomized to one of two modes of screening for ovarian cancer, or a control group. The researchers rigorously pre-specified a protocol in which the primary analysis was a reduction in mortality from ovarian cancer, assessed using a statistical method that assumed the proportional reduction in risk would be the same over the whole period of follow-up.[1]

When the data was eventually analysed after an average follow-up of eleven years, the pre-specified primary analysis did not show a statistically significant benefit, and the authors duly reported this non-significant result as their main conclusion. So why did the *Independent* newspaper have the headline 'Ovarian Cancer Blood Tests Breakthrough: Huge Success of New Testing Method Could Lead to National Screening in Britain'?[2]

We shall come back to whether the results of this massive, and very expensive, study were appropriately interpreted.

*

In the last chapter we saw how poor practice can occur all along the pipeline of statistical stories, which means that if we want the use of statistics to improve, three groups need to act:

- *Producers of statistics*: such as scientists, statisticians, survey companies and industry. They can do statistics better.
- *Communicators*: such as scientific journals, charities, government departments, press officers, journalists and editors. They can communicate statistics better.
- *Audiences*: such as the public, policy-makers and professionals. They can check statistics better.

We can consider in turn what each of these groups might do.

Improving What Is Produced

How can the whole process of science become more reliable? A varied collaboration of prominent researchers have developed a 'reproducibility manifesto', which covers improved research methods and training, promoting the pre-registration of the design and analysis of studies, better reporting of what was actually done, encouraging replication studies, diversifying peer review, and rewarding openness and transparency.[3] Many of these ideas are reflected in the Open Science Framework, a facility that particularly encourages data-sharing and pre-registration of studies[4].

Given the examples in the last chapter, it should come as no surprise that many of the suggestions in the manifesto concern statistical practice, and in particular the appeal to pre-register studies is intended to guard against the kind of behaviour that was so vividly illustrated in the last chapter, where the design, hypotheses and analysis of a study were adapted to the data as it arrived. But it could also be argued that complete pre-specification is not realistic and denies the imagination of the researcher, and that the flexibility to adapt to new data is a positive characteristic. Again, the answer seems to lie in a clear distinction between exploratory and confirmatory studies, always with clear reporting of the sequence of choices that the researchers have made.

Pre-specification of an analysis is not without problems since it may constrain researchers to an analysis that, when the data come in, they realize is inappropriate. For example, the team that carried out the ovarian cancer screening trial planned to include all randomized patients in their analysis, but found that if 'prevalent' cases (those who were found to have ovarian cancer before the trial started) were excluded from the analysis, which might be considered quite a reasonable thing to do, then the multi-modal screening strategy *did* show a significant 20% reduction in mortality from ovarian cancer ($P = 0.02$). In addition, even when including all cases regardless of whether or not they had ovarian cancer at the start of the trial, a significant 23% reduction in mortality in the multi-modal group also appeared in the period between 7 and 14 years after randomization. So issues that may not have been foreseen, such as people who already had ovarian cancer being randomized, and screening taking some time to be effective, prevented the pre-planned primary outcome being significant.

The authors were meticulous in reporting that their

primary analysis did not show a significant result, and ruefully commented that 'the main limitation of this trial was our failure to anticipate the late effect of screening in our statistical design.' This did not stop some of the media interpreting a non-significant result as confirming the null hypothesis, and incorrectly reporting that the study showed that screening did not work at all. The *Independent* headline claiming screening could save thousands of lives, although somewhat bold, perhaps better reflected the conclusions of the study.

Improving Communication

This book has featured some dire media coverage of statistics-based stories. There is no simple way to influence the practice of journalism and the media, particularly in a time when the industry is being challenged due to competition from social media and unregulated online publications, and advertising revenues are dwindling, but statisticians have collaborated on reporting guidelines for media organizations and training programmes for journalists and press officers. The good news is that data journalism is flourishing, and collaborations with journalists can lead to enriched storytelling based on data, featuring appropriate and attractive narratives and visualizations.

There are, however, risks to turning numbers into stories. Traditional storylines need an emotional hit, a strong narrative arc and a well-rounded conclusion; science rarely provides all of these, and so it is tempting to oversimplify and overclaim. Stories should be true to the evidence: its strengths, weaknesses and uncertainties. Ideally stories might say that a drug or another medical intervention is neither good nor bad: it has benefits and side effects, that people might weigh them up in different ways, and quite reasonably come to different conclusions. Journalists seem to shy away from such nuanced narratives but (say by including testimony from people with differing views) a good communicator should be able to make these stories gripping. For example, Christie Aschwanden from *FiveThirtyEight* discussed the statistics about breast screening, and then said that she had decided to avoid the procedure, whereas her smart friend, provided with precisely the same evidence, had made the opposite decision.[5] This neatly asserts the importance of personal values and concerns, while still respecting the statistical evidence.

We could also do better research on how best to improve the communication of statistics. For example, how can we best communicate uncertainty about facts and the future without jeopardizing trust and credibility, and how can our techniques be tailored to audiences with different attitudes and knowledge? These are important and researchable questions. In addition, the dismal level of statistical debate in the UK Brexit referendum campaign suggests the need for research into different ways of communicating how policy decisions might impact society.

Helping to Call Out Poor Practice

Many individuals and groups have a role in identifying poor statistical practice, including referees of papers that have been submitted for publication, those conducting systematic reviews of published evidence, journalists and fact-checking organizations, and individual members of the public.

Uri Simonsohn has been particularly forthright in arguing that referees should be more stringent in making sure that a paper's authors have followed the requirements set down by the journal, can demonstrate that their results are robust and do not depend on arbitrary decisions in the analysis, and require replication if there is any doubt. On the other hand, he suggests referees should be more tolerant of imperfections in the results, which should encourage honest reporting.[6]

But, as someone who has personally refereed hundreds of scientific papers, I would say that it is not always straightforward to identify the problems. Checklists can be useful but can be gamed by authors to make the paper seem reasonable. I have to admit to developing a suspicious 'nose' which starts to sniff for hints that, for example, large numbers of comparisons have been made and only the 'interesting' ones are being reported.

Such a nose should definitely start reacting when a result seems too good to be true, such as when a large effect has been observed in a small sample. A classic example is a much-publicized 2007 study that claimed to show that attractive people had more daughters. A US survey of adolescents had had their 'physical attractiveness' rated on a five-point scale, and fifteen years later, of those who had been rated as 'very attractive' as an adolescent, only 44% of their first-born were boys, as opposed to the standard 52% for all the plainer people (as Arbuthnot showed, on average there are slightly more boys than girls born). This finding is statistically significant, but as Andrew Gelman identified, it is far too large an effect to be plausible, and also occurs only in the 'most attractive' group. There is nothing in the paper that will reveal the total implausibily of this result – external knowledge is required.[7]

Publication Bias

Scientists examine huge numbers of published articles when they are conducting systematic reviews – trying to bring together the literature and synthesize the current state of knowledge. Such an enterprise becomes hopelessly flawed if what is published is a biased subset of the work that has been carried out, say because negative results have not been submitted for publication, or questionable research practices have led to an unjustified excess of significant results.

Statistical techniques have been developed for identifying such publication bias. Suppose we have a set of studies that all set out to test the same null hypothesis that an intervention has no effect. Regardless of the actual experiments conducted, if the intervention really has no effect, it can be proved theoretically that any P-value that tests the null hypothesis is equally likely to take on any value between 0 and 1, and so the P-values from many studies testing the effect should tend to scatter uniformly. Whereas if there really is an effect, the P-values will tend to be skewed towards small values.

The idea of the 'P-curve' is to look at all the actual P-values reported for significant test results – that is, when $P < 0.05$. Two features create suspicion. First, if there is cluster of P-values just below 0.05, it suggests some massaging has been done to tip some of them over this crucial boundary. Second, suppose these significant P-values are not skewed towards 0, but fairly uniformly scattered between 0 and 0.05. Then this is just the pattern that would occur were the null hypothesis true, and the only results being reported as significant were those 1 in 20 that tipped over $P < 0.05$ by luck. Simonsohn and others looked at the published psychological literature which supported the popular idea that giving people an excessive amount of choice led to negative consequences; an analysis of the P-curve suggested there was substantial publication bias and that there was no good evidence for this effect.[8]

Assessing a Statistical Claim or Story

Whether we are journalists, fact-checkers, academics, professionals in government or business or NGOs, or simply members of the public, we are regularly told claims that are based on statistical evidence. Assessing the trustworthiness of statistical claims appears a vital skill for the modern world.

Let us make the bold assumption that all those involved in the collection, analysis and use of statistics are adhering to an ethical framework in which trust is paramount. Onora O'Neill, an eminent philosopher of Kant and an authority on trust, has stressed that people should not seek to be trusted, since that is granted by others, but to demonstrate the *trustworthiness* of their work. O'Neill has provided some perceptive short checklists: for example, trustworthiness requires honesty, competence and reliability. But she also points out that evidence of trustworthiness is required, and this means being *transparent* – not by just dumping masses of data on audiences, but being 'intelligently transparent'.[9] This means that claims based on data need to be:

- *Accessible*: audiences should be able to get at the information.
- *Intelligible*: audiences should be able to understand the information.
- *Assessable*: if they wish to, audiences should be able to check the reliability of the claims.
- *Useable*: audiences should be able to exploit the information for their needs.

But assessing trustworthiness is not a straightforward task. Statisticians and others spend decades learning how to weigh up claims, and coming up with questions that will help identify any flaws. It is not a matter of a simple checklist: experience and a somewhat sceptical attitude may be required. With this caveat, here is a set of questions that tries to encapsulate whatever wisdom there may be in this book. The terms and issues that might be considered for each one are either self-explanatory or covered earlier: I find this list useful, and I hope you might too.

Ten Questions to Ask When Confronted by a Claim Based on Statistical Evidence

HOW TRUSTWORTHY ARE THE NUMBERS?

1. *How rigorously has the study been done?* For example, check for 'internal validity', appropriate design and wording of questions, pre-registration of the protocol, taking a representative sample, using randomization, and making a fair comparison with a control group.

2. *What is the statistical uncertainty / confidence in the findings?* Check margins of error, confidence intervals, statistical

significance, sample size, multiple comparisons, systematic bias.

3. *Is the summary appropriate?* Check appropriate use of averages, variability, relative and absolute risks.

HOW TRUSTWORTHY IS THE SOURCE?

4. *How reliable is the source of the story?* Consider the possibility of a biased source with conflicts of interest, and check publication is independently peer-reviewed. Ask yourself, 'Why does this source want me to hear this story?'

5. *Is the story being spun?* Be aware of the use of framing, emotional appeal through quoting anecdotes about extreme cases, misleading graphs, exaggerated headlines, big-sounding numbers.

6. *What am I not being told?* This is perhaps the most important question of all. Think about cherry-picked results, missing information that would conflict with the story, and lack of independent comment.

HOW TRUSTWORTHY IS THE INTERPRETATION?

7. *How does the claim fit with what else is known?* Consider the context, appropriate comparators, including historical data, and what other studies have shown, ideally in a meta-analysis.

8. *What's the claimed explanation for whatever has been seen?* Vital issues are correlation v. causation, regression to the mean, inappropriate claim that a non-significant result means 'no effect', confounding, attribution, prosecutor's fallacy.

9. *How relevant is the story to the audience?* Think about generalizability, whether the people being studied are a special case, has there been an extrapolation from mice to people.

10. *Is the claimed effect important?* Check whether the magnitude of the effect is practically significant, and be especially wary of claims of 'increased risk'.

Data Ethics

Increasing concern about the potential misuse of personal data, particularly when harvested from social media accounts, has focused attention on the ethical aspects of data science and statistics. While government statisticians are bound by an official code of conduct, the more general discipline of data ethics is still in the development stage.

This book has covered the need for algorithms that affect people to be fair and transparent, the importance of doing honest and reproducible science, and the requirement for trustworthy communication. All these form part of data ethics, and the featured stories have shown the harm of allowing conflicts of interest, or even just overenthusiasm, to distort principled practice. Many other key topics could have featured: privacy and ownership of data, informed consent for its wider use, legal aspects of explanation of algorithms, and so on.

Although statistical science may appear a highly technical subject, it always takes place in the context of a society to which its exponents bear a responsibility. In the near future we can expect that data ethics will form an integral part of statistics training.

An Example of Good Statistical Science

Before the UK general election on 8 June 2017, most polls suggested the Conservatives would have a substantial majority. Minutes after voting stopped at 10pm, a team of statisticians predicted the Conservatives had lost many seats and their overall majority, meaning there would be a hung parliament. This claim was greeted with incredulity. How did the team make this bold prediction, and were they right?

In a book that has tried to celebrate good practice in the art and science of learning from data, and not just dwell on misleading studies, it seems appropriate to finish with a fine example of statistical science.

It may seem a curious question to ask who won the election immediately after it has taken place: after all, we can just sit up all night and wait to hear the results. But it has become part of the theatre of elections that, just minutes after the polls close, pundits are making predictions about what the results will turn out to be. Note that the results are already fixed, but are unknown at this time, and so this is a classic example of the kind of epistemic uncertainty that crops up when considering unemployment rates and other quantities that are 'out there' but just not known.

Let's consider the PPDAC cycle. The Problem is to produce a fast prediction of the results of the UK election within minutes of the close of polling. The team, comprising statisticians David Firth and Jouni Kuha, and psephologist John Curtice, came up with a Plan to conduct exit polls, in which around 200 voters are interviewed leaving each of 144 out of around 40,000 polling stations – crucially these 144 polling stations are the same ones as were used in previous exit polls. The Data comprise the responses from participants being asked not only how they voted but, most importantly, how they voted in the previous election.

The Analysis uses a repertoire of techniques, best seen through the stages in drawing inferences laid out in Chapter 3.

- *Data to Sample*: Since these are exit polls and the respondents are saying what they have done and not what they intend to do, experience suggests the responses should be reasonably accurate measures of what people actually voted at this and previous elections.
- *Sample to Study Population*: A representative sample is taken of those who actually voted in each polling station, so the results from the sample can be used to roughly estimate the change in vote, or 'swing', in that small area.
- *Study Population to Target Population*: Using knowledge of the demographics of each polling station, a regression model is built that attempts to explain how the proportion of people who change their vote between elections depends on the characteristics of the voters in that polling area. In this way the swing does not have to be assumed to be the same throughout the country, but is allowed to vary from area to area – allowing, say, for whether there is a rural or urban population. Then using the estimated regression model, knowledge of the demographics of

each of the 600 or so constituencies, and the votes
cast at the previous election, a prediction of the votes
cast in this election can be made for each individual
constituency, even though most of the constituencies did
not actually have any voters interviewed in the exit poll.
This is essentially the multi-level regression and post-
stratification (MRP) procedure outlined in Chapter 11.

The limited sample means that there is uncertainty about the
coefficients in the regression model, which when scaled up to
the entire voting population, produce probability distributions
of how people voted, and hence a probability of each candidate
getting the maximum number of votes. Adding these up across
all the constituencies gives an expected number of seats, each
of which has accompanying uncertainty (although the margins of error were not reported on the night of the election).[10]

Table 13.1 shows the predictions and the final results in the
June 2017 election. The predicted number of seats is remarkably close, being at most four seats away from the final count
for all parties. The table shows that in the last three UK elections, this sophisticated statistical methodology has had exceptional accuracy. In 2015 they predicted huge losses to the
Liberal Democrats, estimating a fall from 57 seats to 10, and a
prominent Liberal Democrat politician, Paddy Ashdown, said
in a live television interview that he would 'eat his hat' if they
were correct. In fact the Liberal Democrats won only 8 seats.*

---

* There is no record of Paddy Ashdown carrying out his promise, although he is still
teased about it, and I was in a radio interview discussing this poll when a large
chocolate hat was brought in for us all to share.

Year	Seats	Conservative	Labour	Liberal Democrat	Scottish Nationalist	Others
2010	Predicted	307	255	59		29
	Actual	307	258	57		28
2015	Predicted	316	239	10	58	27
	Actual	331	232	8	56	23
2017	Predicted	314	266	14	34	21
	Actual	318	262	12	35	22

Table 13.1
Exit poll predictions of the number of seats won by each party in three recent UK national elections just after close of voting, compared with actual results observed. The predictions are estimates and have accompanying margins of error.

The media only provided an estimate for the predicted number of seats, but a margin of error for the winning total is said to be around 20 seats. Past accuracy has been somewhat better than that, so perhaps the statisticians have been rather lucky. But they deserve their luck: they have beautifully demonstrated how statistical science can lead to powerful conclusions that can impress public and professionals alike. Those audiences have little conception of the complexity of the underlying methods, and that this extraordinary performance is due to meticulous attention to detail throughout the whole problem-solving cycle.

Summary

- Producers, communicators and audiences all have a role in improving the way that statistical science is used in society.
- Producers need to ensure that science is reproducible. To demonstrate trustworthiness, information should be accessible, intelligible, assessable and useable.
- Communicators need to be wary of trying to fit statistical stories into standard narratives.
- Audiences need to call out poor practice by asking questions about the trustworthiness of their numbers, their source and their interpretation.
- When faced with a claim based on statistical evidence, first feel whether it seems plausible.

CHAPTER 14

In Conclusion

To put it bluntly, statistics can be difficult. Although I have tried to tackle underlying issues in this book rather than getting embroiled in technical detail, the narrative has unavoidably had to rely on some challenging concepts. So congratulations for reaching the end.

Rather than trying to boil down the past chapters into a shortlist of pieces of wise advice, I can take advantage of the following ten simple rules for effective statistical practice. These came from a group of senior statisticians who, mirroring this book, are keen to emphasize the non-technical issues which are generally not taught in statistics courses.[1] I have added my own comments. These 'rules' should be fairly self-evident, and rather neatly summarize the issues tackled in this book.

1. *Statistical methods should enable data to answer scientific questions*: Ask 'why am I doing this?', rather than focusing on which particular technique to use.
2. *Signals always come with noise*: It is trying to separate out the two that makes the subject interesting. Variability is inevitable, and probability models are useful as an abstraction.

3. *Plan ahead, really ahead*: This includes the idea of pre-specification in confirmatory experiments – avoiding researcher degrees of freedom
4. *Worry about data quality*: Everything rests on the data.
5. *Statistical analysis is more than a set of computations*: Do not just plug into formulae or run procedures in software, without knowing why you are doing so.
6. *Keep it simple*: The main communication should be as basic as possible – do not show off skills in complex modelling unless they are really necessary.
7. *Provide assessments of variability*: With the warning that margins of error are generally bigger than claimed.
8. *Check your assumptions*: And make clear when this has not been possible.
9. *When possible, replicate!*: Or encourage others to do so.
10. *Make your analysis reproducible*: Others should be able to access your data and code.

Statistical science plays an important role in all our lives, and is constantly changing in response to the increasing quantity and depth of data becoming available. But the study of statistics does not just have an impact on society in general but on individuals in particular. From a purely personal perspective, putting this book together has made me realize how much my life has been enriched by engaging with statistics. I hope that you might feel the same – if not now, then in the future.

Glossary

absolute risk: the proportion of people in a defined group who experience an event of interest within a specified period of time.

adjustment/stratification: inclusion into a regression model of known confounders which are not of direct interest, but are intended to allow a more balanced comparison between groups. The hope is that estimated effects associated with explanatory variables of interest should then be closer to causal effects.

aleatory uncertainty: unavoidable unpredictability about the future, also known as chance, randomness, luck and so on.

algorithm: a rule or formula that takes input variables and produces an output, such as a prediction, a classification, or a probability.

artificial intelligence (AI): computer programs intended to perform a task normally associated with human abilities.

ascertainment bias: when the chance of a person being sampled, or a feature being observed, depends on some background factor, for example when people in the treated arm of a randomized trial get closer supervision than the control group.

average: a generic term for a single representative value for a set of numbers, for example the mean, median or mode.

Bayes factor: the relative support given by a set of data for two alternative hypotheses. For hypotheses H_0 and H_1, and data x, the ratio is $p(x|H_0)/p(x|H_1)$.

Bayesian: the approach to statistical inference in which probability is used not only for aleatory uncertainty, but also epistemic uncertainty about unknown facts. Bayes' theorem is then used to revise these beliefs in the light of new evidence.

Bayes' theorem: a rule of probability that shows how evidence A updates prior beliefs of a proposition B to produce posterior beliefs $p(B|A)$, through the formula $p(B\,|\,A) = \frac{p(A|B)p(B)}{p(A)}$. This is easily proved: since $p(B \text{ AND } A) = p(A \text{ AND } B)$, the multiplication rule of probability means that $p(B|A)p(A) = p(A|B)p(B)$, and dividing each side by $p(A)$ gives the theorem.

Bernoulli distribution: if X is a random variable which takes on the value 1 with probability p, and 0 with probability $1 - p$, it is known as a Bernoulli trial with a Bernoulli distribution. X has mean p and variance $p(1 - p)$.

bias/variance trade-off: when fitting a model to be used for prediction, increasing complexity will eventually lead to a model that has less bias, in the sense that it has greater potential to adapt to details of the underlying process, but more variance, since there is not enough data to be confident about the parameters in the model. These elements need to be traded off in order to avoid over-fitting.

big data: an increasingly anachronistic phrase sometimes characterized by four Vs: a huge Volume of data, a Variety of sources such as images, social media accounts or transactions, a high Velocity of acquisition, and possible lack of Veracity due to its routine collection.

binary data: variables that can only take on two values, often yes/no responses to a question. Can be mathematically represented by a Bernoulli distribution.

binomial distribution: when there are n independent possibilities for an event to occur, each with the same probability, the observed number of events has a binomial distribution. Technically for n independent Bernoulli trials $X_1, X_2 \ldots X_n$, each with probability p of success, their sum $R = X_1 + X_2 + \ldots + X_n$, has a binomial distribution with mean np and variance $np(1-p)$, where $P(R=r) = \binom{n}{r} p^r (1-p)^{n-r}$. The observed proportion R/n has mean p and variance $p(1-p)/n$: R/n can therefore be considered as an estimator of p, with standard error $\sqrt{p(1-p)/n}$.

blinding: when those engaged in a clinical trial do not know what treatment a patient has been given, in order to avoid bias in outcome assessments. Single blinding is when patients do not know what treatment they have been given, double blinding means the people monitoring the patients do not know their treatment, triple blinding is when treatments are labelled say A and B, and the statisticians analysing the data and the committee monitoring the results do not know which corresponds to the new treatment.

Bonferroni correction: a method for adjusting size (Type I error) or confidence intervals to allow for simultaneous testing of multiple hypotheses. Specifically, when testing n hypotheses, for an overall size (Type I error) of α, each hypothesis is tested with size α/n. Equivalently, $100(1 - \alpha/n)\%$ confidence intervals are quoted for each estimated quantity. For example, when testing 10 hypotheses with an overall α of 5%, then P-values would be compared to $0.05/10 = 0.005$, and 99.5% confidence intervals used.

bootstrapping: a way of generating confidence intervals and the distribution of test statistics through resampling the observed data rather than through assuming a probability model for the underlying random variable. A basic bootstrap sample of a data set $x_1, x_2 \ldots x_n$ is a sample of size n with replacement, so that the bootstrap sample will be drawn from the original set of distinct values, but not generally in the same proportions as the original data set.

Brier score: a measure for the accuracy of probabilistic predictions, based on the mean squared prediction error. If $p_1 \ldots p_n$ are the probabilities given to a set of n binary observations $x_1 \ldots x_n$ taking on values 0 and 1, then the Brier score is $\frac{1}{n}\sum_i^n (x_i - p_i)^2$. Essentially a mean-squared-error criterion applied to binary data.

calibration: the requirement for the observed frequencies of events to match those expected by probabilistic predictions. For example, of the occasions when events are given a probability of 0.7, then the events should actually occur roughly 70% of the time.

case-control study: a retrospective study design in which people with a disease or outcome of interest (the cases) are matched with one or more people who do not have the disease (the controls), and the histories of the two groups are compared to see whether there are exposures which systematically differ between the two groups. This design can only estimate relative risks associated with exposures.

categorical variable: a variable that can take on two or more discrete values, which may or may not be ordered.

Central Limit Theorem: the tendency for the sample mean of a set of random variables to have a normal sampling distribution, regardless (with certain exceptions) of the shape of the underlying sampling distribution of the random variable. If n independent observations each have mean μ and variance σ^2, then under broad assumptions their sample mean is an estimator of μ, and has an approximately normal distribution with mean μ, variance σ^2/n, and standard deviation $\sigma/\sqrt{n}$ (also known as the standard error of the estimator).

chi-squared test of association / goodness-of-fit test: a statistical test that indicates the degree of incompatibility of data with an assumed statistical model comprising the null hypothesis, which may be one of lack of association, or some other specfied mathematical form. Specifically, the test compares a set of m observed counts $o_1, o_2 \ldots o_m$ with a set of expected values $e_1, e_2 \ldots e_m$ which have been calculated under the null hypothesis. The simplest version of the test statistic is given as

$$X^2 = \sum_{j=1}^{m} \frac{(o_j - e_j)^2}{e_j}.$$

Under the null hypothesis X^2 will have an approximate chi-squared sampling distribution, enabling an associated P-value to be calculated.

classification tree: a form of classification algorithm in which features are examined in sequence, with the response indicating the next feature to examine, until a classification is made.

confidence interval: an estimated interval within which an unknown parameter may plausibly lie. Based on an observed set of data x, a 95% confidence interval for μ is an interval whose lower limit $L(x)$ and upper limit $U(x)$ has the property that, before observing the data, there is a 95% probability that the random interval $(L(X), U(X))$ contains μ. The Central Limit Theorem, combined with the knowledge that close to 95% of a normal distribution lies between the mean ± 2 standard deviations, means that a common approximation for a 95% confidence interval is the estimate ± 2 standard errors. Suppose we want to find a confidence interval for the difference $\mu_2 - \mu_1$ between two parameters μ_2 and μ_1. If T_1 is an estimator of μ_1 with standard error SE_1, and T_2 is an estimator of μ_2 with standard error SE_2, then $T_2 - T_1$ is an estimator of $\mu_2 - \mu_1$. The variance of the difference between two estimators is the sum of their variances, and so the standard error of $T_2 - T_1$ is given by $\sqrt{SE_1^2 + SE_2^2}$. From this a 95% confidence interval for the difference $\mu_2 - \mu_1$ can be constructed.

confirmatory studies and analyses: rigorous studies ideally done to a pre-specified protocol to confirm or negate hypotheses suggested by exploratory studies and analyses.

confounder: a variable which is associated with both a response and a predictor, and which may explain some of their apparent relationship. For example, the height and weight of children are strongly correlated, but much of this association is explained by the age of the child.

continuous variable: a random variable X that can, at least in principle, take on any value within a specific range. It has a probability density function f such that $P(X \leq x) = \int_{-\infty}^{x} f(t)dt$, and expectation given by $E(X) = \int_{-\infty}^{\infty} x f(x)dx$. The probability of X lying in the interval (A, B) can be calculated using $\int_A^B f(x)\,dx$.

control group: a set of individuals who have not been subject to the exposure of interest, say by randomization.

control limits: pre-specified limits for a random variable which are used in quality control to monitor deviation from an intended standard, say displayed on a funnel plot.

count variables: variables that can take on integer values 0, 1, 2 and so on.

counter-factual: a 'what-if' scenario in which an alternative history of events is considered.

cross-sectional study: when analysis is based solely on the current state of individuals, without any follow-up over time.

cross-validation: a way of assessing the quality of an algorithm for prediction or classification by systematically removing some cases to act as a test set.

cox regression: See **hazard ratio**.

data literacy: the ability to understand the principles behind learning from data, carry out basic data analyses, and critique the quality of claims made on the basis of data.

data science: the study and application of techniques for deriving insights from data, including constructing algorithms for prediction. Traditional statistical science forms part of data science, which also includes a strong element of coding and data management.

deep learning: a machine-learning technique that extends standard artificial neural network models to many layers representing different levels of abstraction, say going from individual pixels of an image through to recognition of objects.

dependent events: when the probability of one event depends on the outcome of another event.

dependent, response or outcome variable: the variable of primary interest that we wish to predict or explain.

epidemiology: the study of the rates of, and reasons for, the occurrence of disease.

epistemic uncertainty: lack of knowledge about facts, numbers or scientific hypotheses.

error matrix: a cross-tabulation of correct and incorrect classifications by an algorithm.

expectation (**mean**): the mean-average of a random variable. It is defined as $\sum xp(x)$ for a discrete random variable X and $\int xp(x)\,dx$ for a continuous random variable. For example, if X is the result of throwing a fair die, then $P(X = x) = \frac{1}{6}$ for $x = 1, 2, 3, 4, 5, 6$, so that $E(X) = \frac{1}{6}(1 + 2 + 3 + 4 + 5 + 6) = 3.5$.

expected frequencies: the numbers of events expected to occur in the future, according to an assumed probability model.

exploratory studies and analyses: initial flexible studies which allow adaptive changes to design and analyses in order to pursue promising leads, and are intended to generate hypotheses to be tested in confirmatory studies.

exposure: a factor whose impact on a disease, death or other medical outcome is of interest, such as an aspect of the environment or behaviour.

external validity: when the conclusions of a study can be generalized to a target group, wider than the immediate population that has been studied. This addresses the relevance of a study.

false discovery rate: when testing multiple hypotheses, the proportion of positive claims that turn out to be false-positives.

false-positive: an incorrect classification of a 'negative' case as a 'positive' case.

feature engineering: in machine learning, the process of reducing the dimensionality of input variables, creating summary measures intended to encapsulate the information in the whole data.

forensic epidemiology: using knowledge about the causes of disease in populations when making judgements about the causes of disease in individuals.

framing: the choice of how to express numbers, which in turn can influence the impression given to audiences.

funnel plot: a plot of a set of observations from different units against a measure of their precision, where units might be institutions, areas or studies. Often two 'funnels' indicate where we would expect 95% and 99.8% of observations to lie, were there really no underlying differences between the units. When the distribution of the observations is approximately normal, the 95% and 99.8% control limits are essentially the mean ± two and three standard errors.

hazard ratio: when analysing survival times, the relative risk, associated with an exposure, of suffering an event in a fixed period of time. A Cox regression is a form of multiple regression when the response variable is a survival time, and the coefficients correspond to log(hazard ratios).

hierarchical modelling: in Bayesian analysis, when the parameters underlying a number of units, say areas or schools, are themselves assumed to be drawn from a common prior distribution. This results in shrinkage of the parameter estimates for individual units towards an overall mean.

hypergeometric distribution: the probability of k successes in n draws, without replacement, from a finite population of size N that contains exactly K objects with that feature, formally given by

$$\frac{\binom{K}{k}\binom{N-K}{n-k}}{\binom{N}{n}}$$

hypothesis testing: a formal procedure for evaluating the support for hypotheses provided by data, generally an amalgam of classic Fisherian tests of a null hypothesis using a P-value, and the Neyman–Pearson structure of null and alternative hypotheses and Type I and Type II errors.

icon arrays: a graphic display of frequencies using a set of small images, say of people.

independent events: A and B are independent if the occurrence of A does not influence the probability of B, so that $p(B|A) = p(B)$, or equivalently $p(B, A) = p(B)p(A)$.

independent variable / predictor: a variable that is fixed by design or observation, and whose association with an outcome variable may be of interest.

induction / inductive inference: the process of learning about general principles from specific examples.

inductive behaviour: a proposal by Jerzy Neyman and Egon Pearson in the 1930s to frame hypothesis testing in terms of decision-making. The ideas of size, power and Type I and Type II errors are remnants.

intention to treat: the principle by which participants in randomized trials are analysed according to whatever intervention they were supposed to get, whether or not they actually received it.

interactions: when multiple explanatory variables combine to produce an effect different from that expected from their individual contributions.

internal validity: when the conclusions of a study truly apply to the population of a study. This addresses the rigour with which a study has been conducted.

inter-quartile range: a measure of the spread of a sample or a population distribution, specifically the distance between the 25th and 75th percentiles. Equivalent to the difference between the 1st and 3rd quartiles.

Law of Large Numbers: the process by which the sample mean of a set of random variables tends towards the population mean.

least-squares: suppose we have a set of n paired numbers, $(x_1, y_1), (x_2, y_2) \ldots (x_n, y_n)$, and $\bar{x}$, s_x are the sample mean and standard deviation of the xs, and $\bar{y}$, s_y are the sample mean and standard deviation of the ys. Then the least-squares regression line is given by

$$\hat{y} = b_0 + b_1 (x - \bar{x}),$$

where

- $\hat{y}$ is the predicted value for the dependent variable y for a specified value of the independent variable x.
- The gradient is $b_1 = \dfrac{\sum_i (y_i - \bar{y})(x_i - \bar{x})}{\sum_i (x_i - \bar{x})^2}$.
- The intercept is $b_0 = \bar{y}$. The least-squares line goes through the centre of gravity $\bar{x}$, $\bar{y}$.
- The ith residual is the difference between the ith observation and its predicted value, $y_i - \hat{y}_i$.
- The adjusted value of the ith observation is the residual added to the intercept, i.e., $y_i - \hat{y}_i + \bar{y}$. It is intended to be the value we would have observed were this an 'average' case, that is with $x = \bar{x}$ rather than $x = x_i$.
- The residual sum of squares (RSS) is the sum of the squares of the residuals, so that RSS $= \sum_{i=1}^{n} (y_i - \hat{y}_i)^2$. The least-squares line is defined as the line that minimizes the residual sum of squares.
- The gradient b_1 and Pearson's correlation coefficient r are related through the formula $b_1 = r s_y / s_x$. So if the standard deviations of the xs and ys are the same, then the gradient is exactly equal to the correlation coefficient.

likelihood: a measure of the evidential support provided by data for particular parameter values. When a probability distribution for a random variable depends on a parameter, say θ, then after observing data x the likelihood for θ is proportional to $p(x|\theta)$.

likelihood ratio: a measure of the relative support that some data provides for two competing hypotheses. For hypotheses H_0 and H_1, the likelihood ratio provided by data x is given by $p(x|H_0)/p(x|H_1)$.

logarithmic scale: The logarithm to base 10 of a positive number x is denoted by $y = \log_{10} x$, or equivalently $x = 10^y$. In statistical analysis, $\log x$ generally denotes the natural logarithm $y = \log_e x$, or equivalently $x = e^y$ where e is the exponential constant 2.718.

logistic regression: a form of multiple regression when the response variable is a proportion, and the coefficients correspond to log(odds ratios). Suppose we observe a series of proportions $y_i = r_i/n_i$, assumed to arise from a binomial variable with underlying probability p_i with a corresponding set of predictor variables $(x_{i1}, x_{i2} \ldots x_{ip})$. The logarithm of the odds of the estimated probability $\hat{p}_i$ is assumed to be a linear regression:

$$\log \frac{\hat{p}_i}{(1 - \hat{p}_i)} = b_0 + b_1 x_{i1} + b_2 x_{i2} + \ldots + b_p x_{ip}.$$

Suppose one of the predictor variables, say x_1, is binary with $x_1 = 0$ corresponding to not being exposed to a potential hazard, and $x_1 = 1$ corresponding to being exposed. Then the coefficient b_1 is a log (odds ratio).

lurking factor: in epidemiology, an exposure that has not been measured but may be a confounder responsible for some of the observed association: for example, when socioeconomic status has not been measured in a study relating diet with disease.

machine learning: procedures for extracting algorithms, say for classification, prediction or clustering, from complex data.

margin of error: after a survey, a plausible range in which a true characteristic of a population may lie. These are generally 95% confidence intervals, which are approximately ± 2 standard errors, but sometimes error-bars are used to represent ± 1 standard error.

mean (of a population): *see* **expectation**

mean (of a sample): suppose we have a set of n data-points, which we label as $x_1, x_2, \ldots, x_n$. Then their sample mean is given by $m = (x_1 + x_2 + \ldots + x_n)/n$, which can be written as $m = \frac{1}{n} \sum_{i=1}^{n} x_i = \bar{x}$. For example, if 3, 2, 1, 0, 1 are the numbers of children reported by 5 people in a sample, then the sample mean is $(3 + 2 + 1 + 0 + 1)/5 = 7/5 = 1.4$.

mean-squared-error (MSE): a measure of performance when predictions $t_1 \ldots t_n$ are made of observations $x_1 \ldots x_n$, given by $\frac{1}{n} \sum_{i=1}^{n} (x_i - t_i)^2$.

median (of a sample): the value mid-way along the ordered set of data-points. If the data-points are put in order, we denote the lowest by $x_{(1)}$, the second lowest by $x_{(2)}$, and so on until the maximum value $x_{(n)}$. If n is odd, then the sample median is the middle value $x_{\left(\frac{n+1}{2}\right)}$; if n is even, then the average of the two 'middle' points is taken as the median.

meta-analysis: a formal statistical method for combining the results from multiple studies.

mode (of a population distribution): the response with the maximum probability of occurring.

mode (of a sample): the most common value in a set of data.

multi-level regression and post-stratification (MRP): a modern development in survey sampling in which fairly small numbers of responders are obtained from many areas. A regression model is then built relating responses to demographic factors, allowing for additional between-area variability using hierarchical modelling. Knowing the demographics of all areas then allows both local and national predictions to be made, with appropriate uncertainty.

multiple linear regression: suppose that for every response y_i there are a set of p predictor variables $(x_{i1}, x_{i2} \ldots x_{ip})$. Then a least-squares multiple linear regression is given by

$$\hat{y}_i = b_0 + b_1(x_{i1} - \overline{x}_1) + b_2(x_{i2} - \overline{x}_2) + \ldots + b_p(x_{ip} - \overline{x}_p),$$

where the coefficients $b_0, b_1 \ldots b_p$ are chosen to minimize the residual sum of squares RSS $= \sum_{i=1}^{n} (y_i - \hat{y}_i)^2$. The intercept b_0 is simply the mean $\overline{y}$, and the formula for the remaining coefficients is complex but easily computed. Note that $b_0 = \overline{y}$ is the predicted value of an observation y whose predictor variables were the averages $(\overline{x}_1, \overline{x}_2 \ldots \overline{x}_p)$, and, just as for a linear regression, an adjusted y_i is given by the residual plus the intercept, or $y_i - \hat{y}_i + \overline{y}$.

multiple testing: when a series of hypothesis tests are carried out, so increasing the chance of at least one false-positive claim (Type 1 error).

normal distribution: X has a normal (Gaussian) distribution with mean μ and variance σ^2 if it has a probability density function

$$f(x) = \frac{1}{\sqrt{2\pi\sigma^2}} e^{-\frac{(x-\mu)^2}{2\sigma^2}}, \text{ for } -\infty \leq x \leq \infty.$$

Then $E(X) = \mu$, $V(X) = \sigma^2$, $SD(X) = \sigma$. The standardized variable $Z = \frac{X-\mu}{\sigma}$ has mean 0 and variance 1, and is said to have a standard normal distribution. We write Φ for the cumulative probability of a standard normal variable Z.

For example, $\Phi(-1) = 0.16$ is the probability of a standard normal variable being less than -1, or equivalently, the probability of a general normal variable being less then one standard deviation below the mean. The $100p\%$ percentile of the standard normal distribution is z_p where $P(Z \le z_p) = p$. Values of Φ are available in standard software or tables, as are percentage points z_p: for example, the 75th percentile of the standard normal distribution is $z_{0.75} = 0.67$.

null hypothesis: a default scientific theory, generally representing the absence of an effect or a finding of interest, which is tested using a P-value. Generally denoted H_0.

objective priors: an attempt to remove the subjective element in Bayesian analysis, by pre-specifying prior distributions that are intended to represent ignorance about parameters, and so let the data speak for itself. No overall procedure for setting such priors has been established.

odds, odds ratios: if the probability of an event is p, the odds of the event is defined by $\frac{p}{(1-p)}$. If the odds of an event in the exposed group is $\frac{p}{(1-p)}$, and the odds in the non-exposed group is $\frac{q}{(1-q)}$: the odds ratio is then given by $\frac{p}{(1-p)} / \frac{q}{(1-q)}$. If p and q are small, then the odds ratio will be close to the relative risk p/q, but odds ratios and relative risks start to differ when the absolute risks are much more than 20%.

one-sided and two-sided tests: a one-sided hypothesis test is used when a null hypothesis specifies that, say, the effect of

a medical treatment is negative. This would only be rejected
by large positive values of a test statistic representing an
estimated treatment effect. A two-sided test would be
appropriate for a null hypothesis that a treatment effect, say,
is exactly zero, and so both positive and negative estimates
would lead to the null being rejected.

one-tailed and two-tailed P-values: those corresponding to one-
sided and two-sided tests.

over-fitting: building a statistical model that is over-adapted to
training data, so that its predictive ability starts to decline.

parameters: the unknown quantities in a statistical model,
generally denoted with Greek letters.

Pearson correlation coefficient: for a set of n paired numbers,
$(x_1, y_1), (x_2, y_2) \ldots (x_n, y_n)$, when $\bar{x}$, s_x are the sample mean
and standard deviation of the xs, and $\bar{y}$, s_y are the sample
mean and standard deviation of the ys, the Pearson correlation
coefficient is given by

$$r = \frac{\sum_{i=1}^{n} (x_i - \bar{x})(y_i - \bar{y})}{\sum_{i=1}^{n}(x_i - \bar{x})^2 \sum_{i=1}^{n}(y_i - \bar{y})^2}.$$

Suppose xs and ys have both been standardized to Z-scores
given by us and vs respectively, so that $u_i = (x_i - \bar{x})/s_x$, and
$v_i = (y_i - \bar{y})/s_y$. Then the Pearson correlation coefficient can be
expressed as $\sum_{i=1}^{n} u_i v_i$, that is the 'cross-product' of the Z-scores.

percentile (of a population): there is, for example, a 70% chance
of drawing a random observation below the 70th percentile.
For a literal population, it is the value below which 70% of the
population lie.

percentile (of a sample): the 70th percentile of a sample, for
example, is the value that is 70% along the ordered data set:

the median is therefore the 50th percentile. Interpolation between points may be necessary.

permutation/randomization test: a form of hypothesis test in which the distribution of the test statistic under the null hypothesis is obtained by permuting the labels of the data, rather than through a detailed statistical model for the random variables. Suppose the null hypothesis is that a 'label', say being male or female, is not associated with an outcome. Randomization tests examine all possible ways in which labels for individual data-points can be rearranged, each of which are equally likely under the null hypothesis. The test statistic for each of these permutations is calculated, and the P-value is given by the proportion that lead to more extreme test statistics than that actually observed.

placebo: a dummy treatment given to the control arm of a randomized clinical trial, such as a sugar pill disguised to look like the treatment being tested.

Poisson distribution: a distribution for a count random variable X for which $P(X = x | \mu) = e^{-\mu} \frac{\mu^x}{x!}$ for $x = 0, 1, 2 \ldots$ Then $E(X) = \mu$ and $V(X) = \mu$.

population: a group from which it is assumed your sample data are drawn, and which provides the probability distribution for a single observation. In a survey this may be a literal population, but when making measurements, or when having all possible data, the population becomes a mathematical idealization.

population distribution: when the population literally exists, the pattern of potential observations in the entire population. It also refers to the probability distribution of a generic random variable.

posterior distribution: in Bayesian analysis, the probability distribution of unknown parameters after taking into account observed data through Bayes' theorem.

power of a test: the probability of correctly rejecting the null hypothesis, given the alternative hypothesis is true. It is one minus the Type II error rate of a statistical test, and is generally denoted by $1 - \beta$.

PPDAC: a proposed structure for the 'data cycle', comprising Problem, Plan, Data collection, Analysis (exploratory or confirmatory) and Conclusions and communication.

practical significance: when a finding is of genuine importance. Large studies may give rise to results that are statistically but not practically significant.

predictive analytics: using data to create algorithms for making predictions.

prior distribution: in Bayesian analysis, the initial probability distribution for the unknown parameters. After observing data, it is revised to the posterior distribution using Bayes' theorem.

probabilistic forecast: a prediction in the form of a probability distribution for a future event, rather than a categorical judgement of what will happen.

probability: the formal mathematical expression of uncertainty. Let $P(A)$ be the probability for an event A. Then the rules of probability are:

1. Bounds: $0 \leq P(A) \leq 1$, with $P(A) = 0$ if A is impossible and $P(A) = 1$ if A is certain.
2. Complement: $P(A) = 1 - P(\text{NOT } A)$.
3. Addition rule: If A and B are mutually exclusive (i.e., one at most can occur), $P(A \text{ OR } B) = P(A) + P(B)$.

4. Multiplication rule: For any events A and B, $P(A \text{ AND } B) = P(A|B)P(B)$, where $P(A|B)$ represents the probability for A given B has occurred. A and B are independent if and only if $P(A|B) = P(A)$, i.e., the occurrence of B does not affect the probability for A. In this case we have $P(A \text{ AND } B) = P(A)$ $P(B)$, the multiplication rule for independent events.

probability distribution: a generic term for a mathematical expression of the chance of a random variable taking on particular values. A random variable X has a probability distribution function defined by $F(x) = P(X \leq x)$, for all $-\infty < x < \infty$, i.e., the probability that X is at most x.

prosecutor's fallacy: when a small probability of the evidence, given innocence, is mistakenly interpreted as the probability of innocence, given the evidence.

prospective cohort study: when a set of individuals are identified, background factors measured, and then they are followed up and relevant outcomes observed. Such studies are lengthy and expensive, and may not identify many rare events.

P-value: a measure of discrepancy between data and a null hypothesis. For a null hypothesis H_0, let T be a statistic for which large values indicate inconsistency with H_0. Suppose we observe a value t. Then a (one-sided) P-value is the probability of observing such an extreme value, were H_0 true, that is $P(T \geq t|H_0)$. If both small and large values of T indicate inconsistency with H_0, then the two-sided P-value is the probability of observing such a large value in either direction. Often the two-sided P-value is simply taken as double the one-sided P-value, while the R software uses the total probability of events which have a lower probability of occurring than that actually observed.

quartiles (of a population): the 25th, 50th and 75th percentiles.

randomized controlled trial (RCT): an experimental design in which people or other units being tested are randomly allocated to different interventions, thus ensuring, up to the play of chance, that the groups are balanced in both known and unknown background factors. If the groups show subsequent differences in outcome, then either the effect must be due to the intervention or a surprising event has occurred, whose probability can be expressed as a P-value.

random match probability: in forensic DNA testing, the probability that a person randomly drawn from a relevant population would match the observed DNA profile that connects a suspect with a crime.

random variable: a quantity assumed to have a probability distribution. Before they are observed, random variables are generally given a capital letter such as X, while observed values are denoted x.

range (of a sample): the maximum minus the minimum, denoted $x_{(n)} - x_{(1)}$.

rate ratio: the relative increase in the expected number of events in a fixed period of time associated with an exposure. A Poisson regression is a form of multiple regression when the response variable is the observed rate, and the coefficients correspond to log(rate ratios).

Receiver Operating Characteristic (ROC) curve: for an algorithm that generates a score, we can choose a particular threshold for the score above which a unit is classified as 'positive'. As this threshold varies, the ROC curve is formed by plotting the resulting sensitivity (true-positive rate) on the y-axis versus one minus specificity (false-positive rate) on the x-axis.

regression coefficient: an estimated parameter in a statistical model, that expresses the strength of relationship between an explanatory variable and an outcome in multiple regression analysis. The coefficient will have a different interpretation depending on whether the outcome variable is a continuous variable (multiple linear regression), a proportion (logistic regression), a count (Poisson regression) or a survival time (Cox regression).

regression to the mean: when a high or low observation is followed by one that is less extreme, through the process of natural variation. It occurs because part of the reason for the initial extreme case was chance, and this is unlikely to repeat to the same extent.

relative risk: if the absolute risk among people who are exposed to something of interest is p, and the absolute risk among people who are not exposed is q, then the relative risk is p/q.

reproducibility crisis: the claim that many published scientific findings are based on work of insufficient quality, so that the results fail to be reproduced by other researchers.

residual: the difference between an observed value and that predicted by a statistical model.

residual error: the generic term for the component of the data that cannot be explained by a statistical model, and so is said to be due to chance variation.

retrospective cohort study: when a set of individuals are identified at a point in the past, and their subsequent outcomes traced up to the present day. Such a study does not require an extended period of follow-up, but is dependent on the appropriate explanatory variables having been measured in the past.

reverse causation: when an association between two variables initially appears to be causal, but could in fact be acting in the opposite direction. For example, people who do not drink alcohol tend to have poorer health outcomes than moderate drinkers, but this is at least partly due to some non-drinkers having given up alcohol due to poor health.

sample distribution: the pattern made by a set of numerical or categorical observations. Also known as the empirical or data distribution.

sample mean: *see* **mean (of a sample)**

sampling distribution: the probability distribution of a statistic.

sensitivity: the proportion of 'positive' cases that are correctly identified by a classifier or test, often termed the true-positive rate. One minus sensitivity is also known as the observed Type II error or false-negative rate.

sequential testing: when a statistical test is repeatedly carried out on accumulating data, thus inflating the chance of a Type I error occurring at some point. A 'significant result' is guaranteed if the process is continued for long enough.

shrinkage: the influence of a prior distribution in Bayesian analysis, in which an estimate tends to be pulled towards either an assumed or an estimated prior mean. This is also known as 'borrowing strength', since, say, estimated rates of disease in a specific geographical area are influenced by rates in other areas.

signal and the noise: the idea that observed data arises from two components: a deterministic signal which we are really interested in, and random noise that comprises the residual error. The challenge of statistical inference is to appropriately identify the two, and not be misled into thinking that noise is actually a signal.

Simpson's paradox: when an apparent relationship reverses its sign when a confounding variable is taken into account.

size of a test: the Type I error rate of a statistical test, generally denoted by α.

skewed distribution: when a sample or population distribution is highly asymmetric, and has a long left- or right-hand tail. This might typically occur for variables such as income and sales of books, when there is extreme inequality. Standard measures (such as means) and standard deviations can be very misleading for such distributions.

Spearman's rank correlation: the rank of an observation is its position in the ordered set, where 'ties' are considered to have the same rank. For example, for the data (3, 2, 1, 0, 1) the ranks are (5, 4, 2.5, 1, 2.5). Spearman's rank correlation is simply the Pearson's correlation when the xs and ys are replaced by their respective ranks.

specificity: the proportion of 'negative' cases that are correctly identified by a classifier or test. One minus specificity is also known as the observed Type I error, or false-positive rate.

standard deviation: the square root of the variance of a sample or distribution. For well-behaved, reasonably symmetric data distributions without long tails, we would expect most of the observations to lie within two sample standard deviations from the sample mean.

standard error: the standard deviation of a sample mean, when considered as a random variable. Suppose $X_1, X_2 \ldots X_n$ are independent and identically distributed random variables drawn from a population distribution with mean μ and standard deviation σ. Then their average $Y = (X_1 + X_2 + \ldots + X_n)/n$ has mean μ and variance σ^2/n. The standard deviation of Y is

$\sigma/\sqrt{n}$, known as the standard error, and estimated by $s/\sqrt{n}$, where s is the sample standard deviation of the observed X's.

statistic: a meaningful number derived from a set of data.

statistical inference: the process of using sample data to learn about unknown parameters underlying a statistical model.

statistical model: a mathematical representation, containing unknown parameters, of the probability distribution of a set of random variables.

statistical science: the discipline of learning about the world from data, typically involving a problem-solving cycle such as PPDAC.

statistical significance: an observed effect is judged to be statistically significant when its P-value corresponding to a null hypothesis is less than some pre-specified level, say 0.05 or 0.001, meaning such an extreme result was unlikely to occur were the null hypothesis, and all other modelling assumptions, to hold.

supervised learning: construction of a classification algorithm based on cases with confirmed membership of classes.

*t***-statistic**: a test statistic used to test a null hypothesis of a parameter being zero, formed by the ratio of an estimate to its standard error. For large samples, values of above 2 or below −2 correspond to a two-sided P-value of 0.05; exact P-values can be obtained from statistical software.

Type I error: when a true null hypothesis is incorrectly rejected in favour of an alternative, so a false-positive claim is made.

Type II error: when an alternative hypothesis is true, but a hypothesis test does not reject the null hypothesis, so the conclusion is a false-negative.

unsupervised learning: identification of classes based on cases with no identified membership, using some form of clustering procedure.

variability: the inevitable differences that occur between measurements or observations, some of which may be explained by known factors, and the remainder attributed to random noise.

variance: for a sample $x_1 \ldots x_n$ with mean $\bar{x}$, this is generally defined as $s^2 = \frac{1}{(n-1)} \sum_{i=1}^{n} (x_i - \bar{x})^2$ (although the denominator can also be n rather than $n - 1$). For a random variable X with mean μ, the variance is $V(X) = E(X - \mu)^2$. The standard deviation is the square root of the variance, so $SD(X) = \sqrt{V(X)}$.

wisdom of crowds: the idea that a summary derived from a group opinion is closer to the truth than the majority of the individuals.

Z-score: a means of standardizing an observation x_i in terms of its distance from the sample mean m expressed in terms of sample standard deviations s, so that $z_i = (x_i - m)/s$. An observation with a Z-score of 3 corresponds to being 3 standard deviations above the mean, which is a fairly extreme outlier. A Z-score can also be defined in terms of a population mean μ and standard deviation σ, in which case $z_i = (x_i - \mu)/\sigma$.

Notes

INTRODUCTION

1. *The Signal and the Noise* by Nate Silver (Penguin, 2012) is an excellent introduction to how statistical science can be applied to making predictions in sport and other domains.

2. The Shipman data is discussed in more detail in D. Spiegelhalter and N. Best, 'Shipman's Statistical Legacy', *Significance* 1:1 (2004), 10–12. All documents for the public inquiry are available from http://webarchive.nationalarchives. gov.uk/20090808155110/http://www.the-shipman-inquiry.org.uk/reports.asp.

3. T. W. Crowther *et al.*, 'Mapping Tree Density at a Global Scale', *Nature* 525 (2015), 201–5.

4. E. J. Evans, *Thatcher and Thatcherism* (Routledge, 2013), p. 30.

5. *Changes to National Accounts: Inclusion of Illegal Drugs and Prostitution in the UK National Accounts* [Internet] (Office for National Statistics, 2014).

6. The UK Office for National Statistics report a variety of measures of wellbeing at https://www.ons.gov.uk/peoplepopulationandcommunity/ wellbeing.

7. N. T. Nikas, D. C. Bordlee and M. Moreira, 'Determination of Death and the Dead Donor Rule: A Survey of the Current Law on Brain Death', *Journal of Medicine and Philosophy* 41:3 (2016), 237–56.

8. J. P. Simmons and U. Simonsohn, 'Power Posing: *P*-Curving the Evidence', *Psychological Science* 28 (2017), 687–93. For a rebuttal, see A. J. C. Cuddy, S. J. Schultz and N. E. Fosse, '*P*-Curving a More Comprehensive Body of Research on Postural Feedback Reveals Clear Evidential Value for Power-Posing Effects: Reply to Simmons and Simonsohn (2017)', *Psychological Science* 29 (2018), 656–66.

9. A primary recommendation of the American Statistical Association is to 'Teach statistics as an investigative process of problem-solving and decision-making'. See https://www.amstat.org/asa/education/

Guidelines-for-Assessment-and-Instruction-in-Statistics-Education-Reports. aspx. The PPDAC cycle was developed in R. J. MacKay and R. W. Oldford, 'Scientific Method, Statistical Method and the Speed of Light', *Statistical Science* 15 (2000), 254–78. It is strongly promoted in the New Zealand schools system, which provides a highly developed education in statistics. See C. J. Wild and M. Pfannkuch, 'Statistical Thinking in Empirical Enquiry', *International Statistical Review* 67 (1999), 223–265, and the online course Data to Insight https://www.futurelearn.com/courses/data-to-insight.

CHAPTER 1: GETTING THINGS IN PROPORTION: CATEGORICAL DATA AND PERCENTAGES

1. See 'History of Scandal', *Daily Telegraph*, 18 July 2001, and D. J. Spiegelhalter *et al.*, 'Commissioned Analysis of Surgical Performance Using Routine Data: Lessons from the Bristol Inquiry', *Journal of the Royal Statistical Society: Series A (Statistics in Society)* 165 (2002), 191–221.
2. Data on outcomes of child heart surgery in the UK can be obtained from http://childrensheartsurgery.info/.
3. See A. Cairo, *The Truthful Art: Data, Charts, and Maps for Communication* (New Riders, 2016), and *The Functional Art: An Introduction to Information Graphics and Visualization* (New Riders, 2012).
4. World Health Organization. Q&A on the carcinogenicity of the consumption of red meat and processed meat is at http://www.who.int/features/qa/cancer-red-meat/en/. 'Bacon, Ham and Sausages Have the Same Cancer Risk as Cigarettes Warn Experts', *Daily Record*, 23 October 2015.
5. This was a favourite observation of Hans Rosling – see next chapter.
6. E. A. Akl *et al.*, 'Using Alternative Statistical Formats for Presenting Risks and Risk Reductions', *Cochrane Database of Systematic Reviews* 3 (2011).
7. 'Statins Can Weaken Muscles and Joints: Cholesterol Drug Raises Risk of Problems by up to 20 per cent', *Mail Online*, 3 June 2013. The original study is I. Mansi *et al.*, 'Statins and Musculoskeletal Conditions, Arthropathies, and Injuries', *JAMA Internal Medicine* 173 (2013), 1318–26.

CHAPTER 2: SUMMARIZING AND COMMUNICATING NUMBERS. LOTS OF NUMBERS

1. F. Galton, '*Vox Populi*', *Nature* (1907); available at https://www.nature.com/articles/075450a0.

2. In the film (https://www.youtube.com/watch?v=n98BhnwWmsc) of our experiment I rather arbitrarily removed 33 highest guesses of 9,999 or more, took logarithms to give a satisfyingly symmetric distribution, took the mean-average of this transformed distribution, and transformed backwards to get the estimate on the original scale. This gave 1,680 as the 'best guess', which turned out to be the closest of all estimates to the true value of 1,616. This process: take logarithms, calculate mean-average, reverse-logarithm of answer, leads to what is known as the geometric mean. This is equivalent to multiplying all the numbers together and, if there are n numbers, taking their nth root.

    The geometric mean is used in the creation of some economic indices, particularly those based on ratios. This is because it has the advantage that it doesn't matter which way up the ratio goes: the cost of oranges could be measured in pounds per orange, or oranges per pound, and it would still end up with the same geometric mean, whereas this arbitrary choice might make a big difference in mean-averages.

3. C. H. Mercer *et al.*, 'Changes in Sexual Attitudes and Lifestyles in Britain through the Life Course and Over Time: Findings from the National Surveys of Sexual Attitudes and Lifestyles (Natsal)', *The Lancet* 382 (2013), 1781–94. For a vivid examination of the sex statistics, see D. Spiegelhalter, *Sex by Numbers* (Wellcome Collection, 2015).

4. A. Cairo, 'Download the Datasaurus: Never Trust Summary Statistics Alone; Always Visualize Your Data', http://www.thefunctionalart.com/2016/08/download-datasaurus-never-trust-summary.html.

5. https://esa.un.org/unpd/wpp/Download/Standard/Population/.

6. ONS popular names is at https://www.ons.gov.uk/peoplepopulationand community/birthsdeathsandmarriages/livebirths/bulletins/babyname senglandandwales/2015.

7. I. D. Hill, 'Statistical Society of London – Royal Statistical Society: The First 100 Years: 1834–1934', *Journal of the Royal Statistical Society: Series A (General)* 147:2 (1984), 130–39.

8. http://www.natsal.ac.uk/media/2102/natsal-infographic.pdf.

9. H. Rosling, *Unveiling the Beauty of Statistics for a Fact-Based World View*, available on www.gapminder.org.

CHAPTER 3: WHY ARE WE LOOKING AT DATA ANYWAY? POPULATIONS AND MEASUREMENT

1. This four-stage structure is stolen from Wayne Oldford.
2. Ipsos MORI, *What the UK Thinks* (2015), https://whatukthinks.org/eu/poll/ipsos-mori-141215.
3. Reported on *More or Less*, 5 October 2018; https://www.bbc.co.uk/programmes/p06n2lmp. The classic demonstration of priming occurs in the UK comedy series *Yes, Prime Minister,* when top civil servant Sir Humphrey Appleby shows how suitable leading questions can result in any answer desired. This example is now used in teaching research methods. https://researchmethodsdataanalysis.blogspot.com/2014/01/leading-questions-yes-prime-minister.html.
4. Video of Vietnam war draft is at: https://www.youtube.com/watch?v=-p5X1FjyD_g – see also http://www.historynet.com/whats-your-number.htm.
5. Details of the Crime Survey of England and Wales and police-recorded crime can be obtained from the Office for National Statistics: https://www.ons.gov.uk/peoplepopulationandcommunity/crimeandjustice.
6. US birth weights are at: http://www.cdc.gov/nchs/data/nvsr/nvsr64/nvsr64_01.pdf.

CHAPTER 4: WHAT CAUSES WHAT?

1. 'Why Going to University Increases Risk of Getting a Brain Tumour', *Mirror Online*, 20 June 2016. The original article is A. R. Khanolkar *et al.*, 'Socioeconomic Position and the Risk of Brain Tumour: A Swedish National Population-Based Cohort Study', *Journal of Epidemiology and Community Health* 70 (2016), 1222–8.
2. T. Vigen, http://www.tylervigen.com/spurious-correlations.
3. 'MRC/BHF Heart Protection Study of Cholesterol Lowering with Simvastatin in 20,536 High-Risk Individuals: A Randomised Placebo-Controlled Trial', *The Lancet* 360 (2002), 7–22.
4. Cholesterol Treatment Trialists' (CTT) Collaborators, 'The Effects of Lowering LDL Cholesterol with Statin Therapy in People at Low Risk of Vascular Disease: Meta-Analysis of Individual Data from 27 Randomised Trials', *The Lancet* 380 (2012), 581–90.
5. Behavioural Insights Team trials are described in http://www.behavioural insights.co.uk/education-and-skills/helping-everyone-reach-their-potential-

new-education-results/ and http://www.behaviouralinsights.co.uk/trial-results/measuring-the-impact-of-body-worn-video-cameras-on-police-behaviour-and-criminal-justice-outcomes/.

6. H. Benson *et al.*, 'Study of the Therapeutic Effects of Intercessory Prayer (STEP) in Cardiac Bypass Patients: A Multicenter Randomized Trial of Uncertainty and Certainty of Receiving Intercessory Prayer', *American Heart Journal* 151 (2006), 934–42.

7. J. Heathcote, 'Why Do Old Men Have Big Ears?', *British Medical Journal* 311 (1995), https://www.bmj.com/content/311/7021/1668. See also, 'Big Ears: They Really Do Grow as We Age', *The Guardian*, 17 July 2013.

8. 'Waitrose Adds £36,000 to House Price', *Daily Mail*, 29 May 2017.

9. 'Fizzy Drinks Make Teenagers Violent', *Daily Telegraph*, 11 October 2011.

10. S. Coren and D. F. Halpern, 'Left-Handedness: A Marker for Decreased Survival Fitness', *Psychological Bulletin* 109 (1991), 90–106. For criticism, see 'Left-Handedness and Life Expectancy', *New England Journal of Medicine* 325 (1991), 1041–3.

11. J. A. Hanley, M. P. Carrieri and D. Serraino, 'Statistical Fallibility and the Longevity of Popes: William Farr Meets Wilhelm Lexis', *International Journal of Epidemiology* 35 (2006), 802–5.

12. J. Howick, P. Glasziou and J. K. Aronson, 'The Evolution of Evidence Hierarchies: What Can Bradford Hill's "Guidelines for Causation" Contribute?', *Journal of the Royal Society of Medicine* 102 (2009), 186–94.

13. Mendelian randomization has been used, for example, to examine the contested issue of whether moderate alcohol consumption has some health benefits; people who have never drunk alcohol tend to have higher mortality rates than people who drink a little, but there is disagreement over whether this is because of the alcohol or because teetotallers are less healthy for other reasons.

One version of a gene is associated with a decreased tolerance for alcohol and so those people inheriting it tend to drink less. Those with and without that version of the gene should be balanced in all other factors, so any systematic differences in their health can be attributed to the gene, just as in a randomized trial. Researchers have found that people with the gene that decreased tolerance to alcohol tend to be healthier, and concluded this means that alcohol is not good for you. But extra assumptions are needed to draw this conclusion, and the debate is not settled. See Y. Cho *et al.*, 'Alcohol Intake and Cardiovascular Risk Factors: A Mendelian Randomisation Study', *Scientific Reports*, 21 December 2015.

CHAPTER 5: MODELLING RELATIONSHIPS USING REGRESSION

1. M. Friendly *et al.*, 'HistData: Data Sets from the History of Statistics and Data Visualization' (2018), https://CRAN.R-project.org/package=HistData.

2. J. Pearl and D. Mackenzie, *The Book of Why: The New Science of Cause and Effect* (Basic Books, 2018), p. 471.

3. For a fascinating discussion of the risk of modelling, see A. Aggarwal *et al.*, 'Model Risk – Daring to Open Up the Black Box', *British Actuarial Journal* 21:2 (2016), 229–96.

4. Essentially we are saying that changes will be correlated with a baseline measure, even if there is really no real change in the underlying process. We can express this mathematically. Suppose I take an observation at random from a population distribution, call it X. Then I take another independent observation from the same distribution, call it Y, and look at their difference: $Y - X$. Then it is a rather remarkable fact that the correlation between their difference, $Y - X$, and the first measurement, X, is $-1/\sqrt{2} = -0.71$, regardless of the form of the underlying population distribution. For example, if a woman has a child, and then her friend has one, and they see how much heavier the friend's baby is by taking the weight of the second minus the weight of the first, then this difference has a correlation of -0.71 with the weight of the first baby. This is because, if the first child is light, we expect the second to be heavier just by chance alone, and so the difference would be positive. And if the first child is heavy, then we expect the difference between the weights to be negative.

5. L. Mountain, 'Safety Cameras: Stealth Tax or Life-Savers?', *Significance* 3 (2006), 111–13.

6. The table below shows the forms of multiple regression used for different types of dependent variable. Each results in a regression coefficient being estimated for each explanatory variable.

Type of dependent variable	Type of regression	Interpretation of coefficient
Continuous variables	Multiple linear	Gradient
Events or proportions	Logistic	Log(odds ratio)
Counts	Poisson	Log(rate ratio)
Length of survival	Cox	Log(hazard ratio)

CHAPTER 6: ALGORITHMS, ANALYTICS AND PREDICTION

1. The *Titanic* data can be downloaded from http://biostat.mc.vanderbilt.edu/wiki/pub/Main/DataSets/titanic3.xls.
2. Verifying probability of precipitation: http://www.cawcr.gov.au/projects/verification/POP3/POP3.html.
3. 'Electoral Precedent', *xkcd*, https://xkcd.com/1122/.
4. http://innovation.uci.edu/2017/08/husky-or-wolf-using-a-black-box-learning-model-to-avoid-adoption-errors/.
5. The use of COMPAS and MMR algorithms is critiqued in C. O'Neil, *Weapons of Math Destruction: How Big Data Increases Inequality and Threatens Democracy* (Penguin, 2016).
6. NHS, Predict: Breast Cancer (2.1): http://www.predict.nhs.uk/predict_v2.1/.

CHAPTER 7: HOW SURE CAN WE BE ABOUT WHAT IS GOING ON? ESTIMATES AND INTERVALS

1. UK labour market statistics, January 2018: https://www.ons.gov.uk/releases/uklabourmarketstatisticsjan2018. Bureau of Labor Statistics, 'Employment Situation Technical Note 2018', https://www.bls.gov/news.release/empsit.tn.htm.

CHAPTER 8: PROBABILITY – THE LANGUAGE OF UNCERTAINTY AND VARIABILITY

1. Consider Game 1. There are many ways of winning, but only one way of losing – throwing four non-sixes in a row. It is therefore easier to find the probability of losing (this is a common trick). The chance of throwing a non-six is $1 - \frac{1}{6} = \frac{5}{6}$ (complement rule), and the chance of throwing four non-sixes in a row is $\frac{5}{6} \times \frac{5}{6} \times \frac{5}{6} \times \frac{5}{6} = \left(\frac{5}{6}\right)^4 = \frac{625}{1296} = 0.48$ (multiplication rule). So the probability of winning is $1 - 0.48 = 0.52$ (complement rule again). Similar reasoning for Game 2 leads to the probability of winning to be $1 - \left(\frac{35}{36}\right)^{24} = 0.49$, showing that Game 1 was slightly more favourable. These rules also show the error in the Chevalier's reasoning – he was adding the probabilities of events that were not mutually exclusive. By his reasoning

the chance of a six when throwing a die 12 times would be 12/6 = 2, which is not very sensible.

2. For discussion and tools for simulation-based methods for teaching statistics, see M. Pfannkuch et al, 'Bootstrapping Students' Understanding of Statistical Inference', TLRI (2013), and K Lock Morgan et al , 'STATKEY: Online Tools for Bootstrap Intervals and Randomization Tests', ICOTS 9 (2014).

3. Comparisons of daily homicide counts with a Poisson distribution: https://www.ons.gov.uk/peoplepopulationandcommunity/crimeandjustice/compendium/focusonviolentcrimeandsexualoffences/yearendingmarch2016/homicide#statistical-interpretation-of-trends-in-homicides.

CHAPTER 9: PUTTING PROBABILITY AND STATISTICS TOGETHER

1. Paul's original blog is here: https://pb204.blogspot.com/2011/10/funnel-plot-of-uk-bowel-cancer.html, and the data can be downloaded from http://pb204.blogspot.co.uk/2011/10/uploads.html.

2. The margin of error is $\pm 2\sqrt{[p(1-p)/n]}$, whose maximum value of $\pm 1/\sqrt{n}$ occurs at $p = 0.5$. Hence the margin of error is at most $\pm 1/\sqrt{n}$, whatever value of the underlying true proportion p.

3. BBC plot of election polls is at: http://www.bbc.co.uk/news/election-2017-39856354.

4. Margins of error for homicide statistics: https://www.ons.gov.uk/peoplepopulationandcommunity/crimeandjustice/compendium/focusonviolentcrimeandsexualoffences/yearendingmarch2016/homicide#statistical-interpretation-of-trends-in-homicides.

CHAPTER 10: ANSWERING QUESTIONS AND CLAIMING DISCOVERIES

1. J. Arbuthnot, 'An Argument for Divine Providence . . .', *Philosophical Transactions* 27 (1710), 186–90.

2. R. A. Fisher, *The Design of Experiments* (Oliver and Boyd, 1935), p. 19.

3. There are $54 \times 53 \times 52 \ldots \times 2 \times 1$ permutations, which is termed '54 factorial' and denoted 54!. This is roughly 2, with 71 zeros following it. Note

that the number of possible ways a deck of 52 cards can be dealt is 52!, and so even if we dealt a million million hands a second, the number of years it would take to work through all possible permutations has 48 zeros after it, whereas the age of the universe is only 14,000,000,000 years. That's why we can be fairly confident that, throughout all card-playing history, no two shuffled decks of cards have ever been in precisely the same order.

4. The dead fish study is described in this poster: http://prefrontal.org/files/posters/Bennett-Salmon-2009.jpg.

5. The CERN announcement of the Higgs boson is at http://cms.web.cern.ch/news/observation-new-particle-mass-125-gev.

6. D. Spiegelhalter, O. Grigg, R. Kinsman and T. Treasure, 'Risk-Adjusted Sequential Probability Ratio Tests: Applications to Bristol, Shipman and Adult Cardiac Surgery', *International Journal for Quality in Health Care* 15 (2003), 7–13.

7. The test statistic has the simple form: SPRT = 0.69 × cumulative observed deaths – cumulative expected deaths. The thresholds are given by $\log((1 - \beta)/\alpha)$.

8. D. Szucs and J. P. A. Ioannidis, 'Empirical Assessment of Published Effect Sizes and Power in the Recent Cognitive Neuroscience and Psychology Literature', *PLOS Biology* 15:3 (2 March 2017), e2000797.

9. J. P. A. Ioannidis, 'Why Most Published Research Findings Are False', *PLOS Medicine* 2:8 (August 2005), e124.

10. C. S. Knott *et al.*, 'All Cause Mortality and the Case for Age Specific Alcohol Consumption Guidelines: Pooled Analyses of up to 10 Population Based Cohorts', *British Medical Journal* 350 (10 February 2015), h384. Reported under the headline, 'Alcohol Has No Health Benefits After All', *The Times*, 11 February 2015.

11. D. J. Benjamin *et al.*, 'Redefine Statistical Significance', *Nature Human Behaviour* 2 (2018), 6–10.

CHAPTER 11: LEARNING FROM EXPERIENCE THE BAYESIAN WAY

1. T. E. King *et al.*, 'Identification of the Remains of King Richard III', *Nature Communications* 5 (2014) 5631.

2. Guidance for communicating likelihood ratios is at: http://enfsi.eu/wp-content/uploads/2016/09/m1_guideline.pdf.

3. For a general article on using Bayes in court, see 'A Formula for Justice', *The Guardian*, 2 October 2011.

4. The formula for this distribution is $60p^2(1-p)^3$, which is technically known as a Beta(3, 4) distribution. With a uniform prior distribution, the posterior distribution for the position of the white ball, having thrown n red balls and r landing to the left of the white, is $\frac{(n+1)!}{r!(n-r)!}p^r(1-p)^{n-r}$, which is a Beta($r+1$, $n-r+1$) distribution.

5. D. K. Park, A. Gelman and J. Bafumi, 'Bayesian Multilevel Estimation with Poststratification: State-Level Estimates from National Polls', *Political Analysis* 12 (2004), 375–85; YouGov results from https://yougov.co.uk/news/2017/06/14/how-we-correctly-called-hung-parliament/.

6. K. Friston, 'The History of the Future of the Bayesian Brain', *Neuroimage* 62:2 (2012), 1230–33.

7. N. Polson and J. Scott, *AIQ: How Artificial Intelligence Works and How We Can Harness Its Power for a Better World* (Penguin, 2018), p. 000.

8. R. E. Kass and A. E. Raftery, 'Bayes Factors', *Journal of the American Statistical Association* 90 (1995), 773–95.

9. J. Cornfield, 'Sequential Trials, Sequential Analysis and the Likelihood Principle', *American Statistician* 20 (1966), 18–23.

CHAPTER 12: HOW THINGS GO WRONG

1. Open Science Collaboration, 'Estimating the Reproducibility of Psychological Science', *Science* 349:6251 (28 August 2015), aac4716.

2. A. Gelman and H. Stern, 'The Difference Between "Significant" and "Not Significant" Is Not Itself Statistically Significant', *American Statistician* 60:4 (November 2006), 328–31.

3. Ronald Fisher, Presidential Address to the first Indian Statistical Congress, 1938, *Sankhyā* 4(1938), 14–17.

4. See 'The Reinhart and Rogoff Controversy: A Summing Up', *New Yorker*, 26 April 2013.

5. 'AXA Rosenberg Finds Coding Error in Risk Program', *Reuters*, 24 April 2010.

6. The Harkonen story is covered in 'The Press-Release Conviction of a Biotech CEO and its Impact on Scientific Research', *Washington Post*, 13 September 2013.

7. D. Fanelli, 'How Many Scientists Fabricate and Falsify Research? A Systematic Review and Meta-Analysis of Survey Data', *PLOS ONE* 4:5 (29 May 2009), e5738.

8. U. Simonsohn, 'Just Post It: The Lesson from Two Cases of Fabricated Data Detected by Statistics Alone', *Psychological Science* 24:10 (October 2013), 1875–88.

9. J. P. Simmons, L. D. Nelson and U. Simonsohn, 'False-Positive Psychology: Undisclosed Flexibility in Data Collection and Analysis Allows Presenting Anything as Significant', *Psychological Science* 22:11 (November 2011), 1359–66.

10. L. K. John, G. Loewenstein and D. Prelec, 'Measuring the Prevalence of Questionable Research Practices with Incentives for Truth Telling', *Psychological Science* 23:5 (May 2012), 524–32.

11. D. Spiegelhalter, 'Trust in Numbers', *Journal of the Royal Statistical Society: Series A (Statistics in Society)* 180:4 (2017), 948–65.

12. P. Sumner *et al.*, 'The Association Between Exaggeration in Health Related Science News and Academic Press Releases: Retrospective Observational Study', *British Medical Journal* 349 (10 December 2014), g7015.

13. 'Nine in 10 People Carry Gene Which Increases Chance of High Blood Pressure', *Daily Telegraph*, 15 February 2010.

14. 'Why Binge Watching Your TV Box-Sets Could Kill You', *Daily Telegraph*, 25 July 2016.

15. Bem's quote is taken from 'Daryl Bem Proved ESP Is Real: Which Means Science Is Broken', *Slate*, 17 May 2017.

CHAPTER 13: HOW WE CAN DO STATISTICS BETTER

1. I. J. Jacobs *et al.*, 'Ovarian Cancer Screening and Mortality in the UK Collaborative Trial of Ovarian Cancer Screening (UKCTOCS): A Randomised Controlled Trial', *The Lancet* 387:10022 (5 March 2016), 945–56.

2. 'Ovarian Cancer Blood Tests Breakthrough: Huge Success of New Testing Method Could Lead to National Screening in Britain', *Independent*, 5 May 2015.

3. M. R. Munafò *et al.*, 'A Manifesto for Reproducible Science', *Nature Human Behaviour* 1 (2017), a0021.

4. Open Science Framework: https://osf.io/.

5. The Aschwanden story is from 'Science Won't Settle the Mammogram Debate', *FiveThirtyEight*, 20 October 2015.

6. J. P. Simmons, L. D. Nelson and U. Simonsohn, 'False-Positive Psychology: Undisclosed Flexibility in Data Collection and Analysis Allows Presenting Anything as Significant', *Psychological Science* 22:11 (November 2011), 1359–66.

7. A. Gelman and D. Weakliem, 'Of Beauty, Sex and Power', *American Scientist* 97:4 (2009), 310–16.

8. U. Simonsohn, L. D. Nelson and J. P. Simmons, 'P-Curve and Effect Size: Correcting for Publication Bias Using Only Significant Results', *Perspectives on Psychological Science* 9:6 (November 2014), 666–81.

9. For more on intelligent openness, see Royal Society, *Science as an Open Enterprise* (2012). Onora O'Neill's perspectives on trustworthiness are brilliantly explained in her TedX talk 'What We Don't Understand About Trust' (June 2013).

10. The methodology for the exit polls has been explained by David Firth at https://warwick.ac.uk/fac/sci/statistics/staff/academic-research/firth/exit-poll-explainer/.

CHAPTER 14: IN CONCLUSION

1. R. E. Kass *et al.*, 'Ten Simple Rules for Effective Statistical Practice', *PLOS Computational Biology* 12:6 (9 June 2016), e1004961.

Index

Locators in *italics* refer to figures and tables

Photo credit: Royal Statistical Society

Sir David John Spiegelhalter is a statistician and chair of the Winton Centre for Risk and Evidence Communication in the Statistical Laboratory at the University of Cambridge. He has served as the president of the Royal Statistical Society and has been knighted for his services to statistics. He lives in Cambridge, UK.